Carnival in the Countryside

IOWA AND THE MIDWEST EXPERIENCE
Series Editor William B. Friedricks
Iowa History Center at Simpson College

Carnival in the Countryside

THE HISTORY OF THE IOWA STATE FAIR

by Chris Rasmussen

University of Iowa Press · Iowa City

University of Iowa Press, Iowa City 52242
Copyright © 2015 by the University of Iowa Press
www.uiowapress.org
Printed in the United States of America

Design by Omega Clay

The University of Iowa Press is a member of Green Press Initiative
and is committed to preserving natural resources.

Printed on acid-free paper

Library of Congress Cataloging-in-Publication Data
Rasmussen, Chris (Chris Allen)
Carnival in the countryside : the history of the Iowa State Fair /
by Chris Rasmussen.
pages   cm
Includes bibliographical references and index.
ISBN 978-1-60938-357-2 (pbk), ISBN 978-1-60938-358-9 (ebk)
1. Iowa State Fair—History. 2. Agricultural
exhibitions—Iowa—History. 3. Country life—Iowa—History.
4. Iowa—Social life and customs. 5. Iowa—Rural conditions.
I. Title.
S555.I8R37   2015
630.74'777—dc23   2015005559

Contents

ACKNOWLEDGMENTS

When I was a graduate student in history, I chose to write about the Iowa State Fair because the fair enabled me to address a wide range of topics, including agriculture, popular culture, film, and art. I also wrote about the fair because, as an Iowan, I wanted to fill in an important chapter in Iowa's history. And, finally, I wrote about the fair because I thought it would be just plain fun, and it was. Many years later, I still find the fair both educational and entertaining.

Catherine Cocks and everyone at the University of Iowa Press and Professor Bill Friedricks offered timely encouragement and excellent advice, and gave me the opportunity to complete a project I began long ago. Marvin Bergman, editor of the *Annals of Iowa*, who has done so much for Iowa's history and all who study it, has offered great advice to me. Margery Tippie copyedited the manuscript with extraordinary attention to detail.

At Rutgers University, Jackson Lears, Paul G. E. Clemens, and Suzanne Lebsock all offered valuable suggestions and encouragement.

I spent a year camped at the State Historical Society of Iowa and in Parks Library at Iowa State University. Archivists at both institutions proved indispensable in locating sources on the fair. I especially want to thank Becki Plunkett at the SHSI and Becky Jordan at Iowa State, both of whom were resourceful and a pleasure to work with. During my year conducting research in Iowa, Lowell Soike and John Zeller were dependable sources of assistance and camaraderie.

Jennifer M. Jones has always been supportive and loving, and has never waned in her belief in me as a historian.

This book is dedicated to my dad, Joel, and to the memory of my mom, Betty. My dad raced stock cars at the fair. My mom went for the country music. Iowans through and through, the fair was a big part of their lives. This book is for them.

Carnival in the Countryside

Introduction

The progress of Iowa might almost be recorded by a history of state fairs. The altering exhibits from year to year, the slow but steady introduction of new features, the growth of the experimental novelty of one year into a staple of a few years later, are a record of the development of the state and its people.
DANTE M. PIERCE
"The State Fair and Its Record of Progress"
The Homestead (1923)

More than a century and a half after its founding, the Iowa State Fair is Iowa's central institution, event, and symbol. New Jersey has the Shore. Kentucky has the Derby. Iowa has the Fair. During its annual run, the fair attracts hundreds of thousands of visitors, who make the pilgrimage to the fairgrounds to see the iconic Butter Cow, ride the Old Mill, walk through the livestock barns, and people-watch. The fair has become not only an icon of Iowa but a slice of Americana, furnishing a picturesque backdrop for presidential candidates, who are obliged to demonstrate their common touch by gobbling down corn dogs and marveling at the victor of the Big Boar contest. The humble Iowa State Fairground ranks alongside the Great Pyramids at Giza and the Taj Mahal in the best-selling travel guide *1,000 Places to See Before You Die*.[1]

The fair today displays the astonishing advances in agriculture and technology over the past century and a half, yet also remains reassuringly familiar. Who could have foreseen driverless tractors, or an Iowa State Fair Food Finder app? Yet today's fair would be recognizable to those Iowans who attended the first fair in 1854, filled as it still is with exhibits of livestock, crops, and crafts; enlivened by shows and games; and thronged with fairgoers ambling about, eating, perusing the fair's exhibits, watching the passing crowds, and just plain having a good time. Iowans have always attended the fair for many reasons—to learn, to compete, to play, to buy, to sell. And they attend it for another reason as well—because they always have. What would August be without the fair?

As historian Karal Ann Marling has aptly observed, "Fairs are our central cultural institution" in the Midwest, and the annual Iowa State Fair remains the nearest thing to a microcosm of Iowa's society and culture. During its annual run, the fair is the very heart of the state, Iowa's most renowned and iconic institution. The fair's Grand Concourse is Iowa's Broadway, its Champs-Elysées. The fair beckons to Iowans as an annual excursion, competition, institution, icon, livestock show, 4-H convention, bazaar, and carnival. It celebrates their achievements in stock raising, farming, and crafts, but its exhibits and competitions are far more than contests for blue ribbons. Collectively they represent an image of Iowa as a cornucopia of productivity and prosperity.[2]

While the fair offers a microcosm of Iowa, it is also a place unto itself. Passing through the fairgrounds gates and being swept along by the river of humanity flowing through its streets and sidewalks, one cannot escape the heady sensation of entering into a time and place apart, one that bombards the senses with dizzying sights, blaring sounds, and appetizing aromas, all beckoning Iowans to set aside their workaday worries. On the fairgrounds many of the rules that govern the other fifty-one weeks of the year seem temporarily to be suspended.

* * *

To understand how the Iowa State Fair came to have these contradictory meanings, we need to consider its long history. Created to promote the adoption of scientific agriculture and foster economic development, state and county fairs became vital institutions in the nineteenth-century United States, and fairgrounds dotted the landscape. American agricultural fairs originated in the Berkshire Mountains of New England but attained their greatest popularity and influence in the Midwest, which seemed destined to become a prosperous agricultural region. Midwestern fairs quickly gained a size and significance generally unmatched elsewhere in the United States. The region's widely dispersed farmers understandably craved an opportunity to exchange ideas, purchase machinery and other goods, and enjoy a respite from the isolation and hard work of farming. Fairs quickly became an annual fall ritual, marking the end of the cycle of planting and harvesting, and spread across the frontier Midwest. Michigan hosted its first state fair in 1849, followed by Ohio (1850), Wisconsin (1851), Indiana (1852), Illinois (1853), Iowa (1854), and Minnesota (1859).

The men who created the Iowa State Fair in 1854 were unabashed optimists and promoters of the state's economy and image, yet even they could scarcely have predicted the fair's long-lived success, considering its modest beginnings. In October 1854, thousands of Iowans hitched up their wagons and journeyed to Fairfield to attend the first Iowa State Fair. As they neared the fairgrounds, they found enterprising showmen, merchants, and gamblers lining the roads, eager to cash in on the vast throng of fairgoers. On the six-acre fairgrounds, hastily constructed buildings shielded the fair's exhibits of crops, machines, cooking, crafts, and paintings from the sun and rain, if not the dust. Estimates of the fair's crowd ranged as high as ten thousand people, making it easily the largest gathering in the brief history of the frontier state. The first fair began with a stirring address on economic development and the dignity of farming and concluded two days later with crowds cheering lustily for a wildly popular contest, "ladies' equestrianism," in which young women on horseback thrilled the crowd at the fair's racetrack. When the fair concluded, organizers, patrons, and journalists alike hailed the exhibition as a resounding success and a landmark event in the state's history.

The men who ran the state fair in the nineteenth century proudly called themselves "agriculturists," who sought to promote economic growth by persuading farmers to adopt scientific agricultural techniques. Most of the men who ran the Iowa State Fair were prominent businessmen, politicians, livestock breeders, implement dealers, and landowners, who had a personal stake in the state's economic development, and they shared fundamental assumptions about its economy and its future. In the language of the day, agriculturists were "boosters," eager to build and promote the economy of their locale and their state. They were genuinely interested in fostering scientific agriculture, but their overarching goal was to create a diversified economy by encouraging manufacturing and other businesses. These boosters sought to make Iowa economically self-sufficient by developing its industrial capacity, so the state would not remain dependent on manufactured goods from Chicago and eastern cities.

Agriculturists promoted economic growth by encouraging farmers to produce staple cash crops, raise purebred livestock, utilize machinery, and tailor their production to the demands of the market. The men who ran the fair measured the progress of their state by tallying its growing population, economy, crops, and head of livestock, along with the fair's growing exhibits, attendance, and receipts. They sought not only to foster economic growth but to promote an image of prosperity in the Midwest, and they considered the fair's exhibits of livestock, crops, farm machinery, handicrafts, and fine arts an advertisement for Iowa's products, prosperity, and promise. Ultimately, they believed, agricultural prosperity would lay the foundation for a distinctive midwestern culture.[3]

Today we too seldom ponder the links between culture and agriculture, and may even consider them virtual opposites: culture, especially when it denotes the arts, learning, and refinement, seems to reside in cities, while agriculture necessarily remains down on the farm, far removed from centers of urbanity and sophistication. But nineteenth-century Americans understood that culture and agriculture were inextricably entwined. In his *American Dictionary of the English Language*, published in 1828, Noah Webster defined "culture" as "the act of tilling and preparing the earth for crops; cultivation; the application of labor, or other means of improvement."[4] Literature, art, and other endeavors that we usually associate with culture were conspicuously absent from Webster's definition. But another Webster, not the lexicographer, defined the relationship between agriculture and culture in a phrase that resonated with American farmers for decades. "When tillage begins, other arts follow," Massachusetts senator Daniel Webster declared in his widely quoted "Remarks on Agriculture" in 1840. "The farmers," he declared, "therefore are the founders of civilization." The Jeffersonian faith in the virtues of tilling the soil, encapsulated in Webster's stirring phrase, remained a powerful current in nineteenth-century American thought, and small-scale, independent producers, especially farmers, were commonly hailed as the embodiment of the republic's virtue and democracy.[5]

When Thomas Jefferson gazed westward and toward the future he envisioned the vast land west of the Appalachians as an "empire for liberty," where independent farmers and agrarian virtue could flourish, untainted by the corrupting influences of urbanization and industrialization. Jefferson was not alone in foreseeing a glorious future for the West. Many settlers hoped that the region's agricultural bounty would give rise to a prosperous and relatively egalitarian society, and they predicted that the "Middle West," as they would later call it, would eventually develop its own distinctive civilization. Tillage had begun, and other arts would surely follow, culminating in a regional economy and culture more prosperous than New England's and more egalitarian than the South's.

The annual state was created to promote economic development, encourage agricultural progress, and ultimately to foster the creation of an indigenous midwestern culture. The fair displayed not only Iowans' accomplishments over the past year but also measured the progress they had made since the outset of the pioneer era. Nineteenth-century Americans commonly thought of the settlement and development of the western frontier as a series of stages, in which agriculture lay at the root of human civilization. Explorers, trappers, and other forerunners of civilization ventured into the wilderness, drove away the Indians, and blazed trails for the pioneering farmers, who cleared trees, broke the prairie, and planted crops, thus planting the beginning of civilization itself. Settlers imposed a tidy grid of farmsteads and townships on the prairie's flowing expanse and built towns, home to merchants and manufacturers. Paths became well-traveled roads, and railroad tracks and telegraph lines spanned the prairie's vast stretches to link far-flung communities. Small towns no doubt sometimes felt like isolated "island communities," in historian Robert Wiebe's famous phrase, but they were linked economically, enmeshed in webs of trade, transportation, and communication. Agriculturists played a large role in creating these webs, and state and county fairs promoted and accelerated the Midwest's economic growth and its integration into the nation's economy.[6]

Hailed as the nation's most prosperous and dynamic region in the late nineteenth century, the Midwest's future seemed promising. But agriculturists were keenly aware that the region's economy was built atop agriculture, and that farming's status was precarious. Although the fair trumpeted the dignity and importance of agriculture, farming's appeal seemed only to dwindle throughout the nineteenth and early twentieth centuries, as industrialization, urbanization, and the advent of an economy geared increasingly toward consumption transformed American society and the growth of factories and cities undermined the Jeffersonian faith in the virtue of agriculture and the status of producers, especially farmers. As the United States rapidly became an urban, industrial nation, the countryside seemed to lag behind economically and culturally. While agriculturists boasted of Iowa's steadily growing economy and population, they fretted that the growth of cities and factories was rapidly eroding farmers' prosperity and status.

Midwesterners discussed their region's progress and prospects in newspaper editorials, campaign speeches, and conversations between neighbors, but nowhere was this discussion more apparent than at the region's state and county fairs. Because fairs encompassed exhibits ranging from pigs to oil paintings, they were widely considered a barometer of a state's or county's economic and cultural development, and they prompted considerable debate over the connection between the Midwest's agricultural economy and its society and culture. When agriculturists charted the fair's growth or fairgoers gazed at exhibits of livestock, handicrafts, or art, they were implicitly taking the measure of their state's economy, society, and culture.

State and county fairs were not only exhibitions of scientific agriculture and economic progress but festive occasions as well. Fairs in Europe had mingled serious purposes, such as commerce and education, with festivity—in which, in this alchemy of seriousness and fun, lies fairs' very essence. In the Middle Ages, European peasants and merchants flocked to fairs, which offered seasonal marketplaces and played an indispensable role in the growth of trade networks. But minstrels, puppeteers, players, and acrobats also performed at these fairs, which became renowned as occasions for amusement. Centuries later, traveling showmen, those modern-day minstrels, pitched their tents at American fairs. Initially, the Iowa State Fair kept most of these showmen outside the fairgrounds—hence the name "sideshows"—but entertainments undeniably attracted people to the fair. As a result, the fair began booking commercial entertainments in the 1870s, and entertainments soon received top billing in the fair's publicity.

The show business became big business in the late nineteenth and early twentieth centuries, and fairs became important venues for carnival companies, grandstand shows, and other amusements. As entertainments became more prominent at the fair, "fair men," who proudly considered themselves members of the outdoor amusement industry, replaced agriculturists at the helm of state and county fairs, and sideshows and carnivals supplanted agricultural exhibits as the fairs' main draw. Critics grumbled that the fair was being diverted from its true purpose, but fair men countered that Iowans would no longer travel to the fair simply to look at livestock and crop displays, and they warned that the fair would go bankrupt if it did not attract patrons by offering up-to-date rides, games, concerts, and other entertainments. They insisted that hard-working farm families, like all Americans, needed leisure and amusement, and that greater access to entertainment would help sustain farming, not undermine it.

As commercial amusements became more prominent at state and county fairs in the late nineteenth and early twentieth centuries, some Iowans embraced entertainments, others tolerated them, and more than a few despised them and sought to bar them from the fair. Some agriculturists and fairgoers acknowledged that shows and games were an indispensable part of the fair, without which it would fail to attract a crowd and just plain go broke. Others considered amusements an expedient, necessary to entice Iowans to the fair's legitimate agricultural exhibits. But staunch opponents of amusements accused shows and games of corrupting the fair, and insisted that the exhibition could succeed without them.

The growth of commercial amusements, such as nickelodeons, vaudeville shows, and amusement parks, is typically associated with urban life, but the outdoor entertainment business transformed leisure and entertainment in the nation's small towns and rural areas as well. In the nineteenth and early twentieth centuries, most Americans lived in the countryside or in small towns, and the outdoor amusement industry catered to rural customers at state and county fairs across the land. The growing prominence of commercial entertainments at state and county fairs provoked debates that differed in key respects from the worries that commonly arose about the amusement business in the nation's cities. Critics of entertainments everywhere fretted that commercial amusements appealed primarily to poorer and less-educated patrons and that they debased Americans' taste and morals. In the Midwest, however, critics also charged that sideshows, carnivals, and grandstand shows threatened to undercut the values and stability of a society built atop agricultural productivity. The fair's exhibits of livestock and crops were commonly seen as embodying rural virtues, hard work, and productivity, while the fair's entertainments signaled the growing influence of urbanization, consumption, and leisure. Entertainments were popular and indispensable to the fair's success, yet their presence on the fairgrounds prompted hand-wringing that the fair was drifting away from its original purpose and that Americans no longer respected farming. For decades, sideshows were a lightning rod for complaints that the fair had forsaken its legitimate purpose, and showmen were blamed for corrupting rural youth and rendering them dissatisfied with farm life by luring them with the temptations of an urbanized, pleasure-seeking society. How you gonna keep 'em down on the farm after they've seen the fair?[7]

When Iowans considered the relationship between agricultural exhibits and entertainments at the fair, they were not merely debating whether it was more important for the fair to build a new hog barn or book a Wild West show. The debate over fairs' educational exhibits and entertainments was more than a squabble over whether peeking under the flap of a tent show to glimpse a belly dancer or a bearded lady would corrupt guileless farm boys, it was a response to momentous changes in America's economy and culture in the late nineteenth and early twentieth centuries. Because agriculture was Iowa's principal industry and the bedrock of the state's culture, any sign that the annual state fair was drifting away from its agricultural mission prompted concern among some Iowans. They considered the state fair a measure of the state's development and a vehicle for promoting agricultural progress. If the annual agricultural exhibition became little more than a carnival, they warned, the fair would contribute to the state's deterioration rather than its development. As a result, the fair's history, and the considerable controversy that often swirled around it, reveals a great deal about agriculture, popular culture, the development of the Midwest, and the region's place in the nation.

When agriculturists, journalists, and fairgoers defended the inclusion of amusements at the fair or complained that showmen had hijacked and corrupted the annual exhibition, they were doing more than debating the propriety of allowing dubious shows and games on the fairgrounds. Their long-running debate over the fair's role was part of a much larger conversation over the progress and prospects of midwestern—and American—society and culture. As historian Jackson Lears has suggested, Americans in the late nineteenth and early twentieth centuries yearned for signs that their nation was undergoing a regeneration of sorts after the cataclysm of the Civil War and amid the momentous changes of the war's aftermath. Manufacturers and businessmen sought to transform America's economy into an industrial colossus and consumer cornucopia. Showmen tantalized patrons with promises of sheer enjoyment and personal liberation. Farmers strove to revive the Jeffersonian faith in the virtues of agriculture. These visions of America's future, compatible or not, could all be found in profusion on the fairgrounds. The history of fair—like the history of Iowa and the Midwest—was neither a straightforward tale of progress or decline, but of people continually gauging the benefits and trade-offs of bewilderingly rapid economic, social, and cultural change. Debates over the state fair's educational exhibits and its entertainments attest to the difficulties that a predominantly rural state, renowned for its agricultural productivity, experienced as the United States became an urban, industrial nation.[8]

Concern about agriculture's precarious status became especially urgent at the outset of the twentieth century, as Americans confronted the realization that many women and children were especially dissatisfied with country life. Farmers' geographic and cultural isolation and the lack of modern conveniences on most farmsteads stoked resentments over the lack of opportunities and amenities available to rural Americans. The recognition that many rural Americans, particularly women and children, found farm life unappealing sparked a concerted public effort in the early twentieth century to improve living conditions in the countryside. University-trained agricultural scientists and home economists launched programs to educate the state's farmers and enlisted state and county fairs to disseminate information and publicize their efforts.

The state fair's educational role was rejuvenated in the 1920s, as professors of agricultural science and home economics from Iowa State College transformed the fair into an annual showcase for 4-H clubs, agricultural and home economics extension projects, and other government-sponsored efforts to improve and sustain rural life. As a result of their efforts, the fair was hailed once again as a crucial institution for preserving and improving rural life in the Midwest. These new agricultural and home economics exhibits were designed not only to educate farmers, especially farm women and children, but also to promote an appealing image of farm life as modern and satisfying, in order to refute the perception that farm women and children yearned to move to town. Paradoxically, while the fair's 4-H and extension exhibits proclaimed the dignity of rural life, they primarily sought to integrate farmers into America's growing consumer society and to diminish the differences between the countryside and the city.[9]

Midwesterners' worries about the viability of rural life were part of their broader concern about the prospects and image of their entire region. In the early twentieth century, the Midwest, a region whose future had formerly appeared so promising, seemed to lag behind the East, as though it were stuck in a rut. By the 1920s, many Americans, and even some midwesterners, disparaged the region as a cultural wasteland, a monotonous landscape inhabited by disgruntled clodhoppers too inert to move to town and small-town residents itching to escape to the big city. The plight of Carol Kennicott, suffocating in Gopher Prairie in Sinclair Lewis's novel *Main Street* (1920), and the conformism and glib boosterism skewered in *Babbitt* (1922), personified the region's stultifying culture in the minds of many Americans.

While critics in the 1920s and 1930s subjected the Midwest to withering scorn as a rural backwater, some midwestern artists and writers responded by creating and extolling a distinct regional culture, one built atop the region's prodigious fertility and productivity. Midwesterners had insisted for decades that their region not only offered fertile ground for farmers but that agricultural productivity would ultimately contribute to the creation of an indigenous regional culture. Even though farming's status had waned, Jeffersonian rhetoric about the dignity of agriculture remained extremely influential well into the twentieth century. The prophesied emergence of an indigenous midwestern culture seemed to be fulfilled when regionalist writers and painters gained renown nationwide in the 1930s, and Iowans figured prominently in the rise of regionalist literature and art. Iowa, in the words of cultural geographer James Shortridge, sets "the standard by which Middle-westernness is measured." Iowa remained more distinctly rural than its neighbors, and its largest city, Des Moines, never became a metropolis on the scale of Chicago, St. Louis, or Cleveland.

As Iowa's central cultural institution, the state fair fittingly, almost inevitably, became both a subject and venue for regionalist writers' and painters' efforts to establish a midwestern culture. In 1932 the fair provided the setting for Phil Stong's best-selling novel, *State Fair*, and for Henry King's successful film the following year. *State Fair* tells the tale of the Frake family's pilgrimage to the fair, during which Abel and Melissa Frake enter the fair's hog and pickle contests, while their children, Wayne and Margy, plunge headlong into romantic liaisons with city dwellers. Stong, an Iowa native, bristled at the stereotype of the Midwest as culturally backward, and *State Fair* depicts farmers as prosperous and content, not downtrodden and bitter. After the novel surprisingly became a runaway success, Stong boasted that he had deliberately sought to launch a "regionalist" school of fiction as a rejoinder to the satires of Sinclair Lewis and the bleak depiction of rural life in Erskine Caldwell's *Tobacco Road*. *State Fair* reassured its readers that farmers remained prosperous and optimistic, and that the temptations of the fair and the city would not entice rural youth away from the farmstead.

While Phil Stong hunched over his typewriter, Grant Wood stood before his easel, painting the most indelible images of the Midwest ever created. An unknown artist, Wood first gained attention for his paintings by entering them in the Iowa State Fair's Art Salon in 1929. Over the next two years, Wood became one of the most famous artists in the United States, and *American Gothic* became an icon nearly as recognizable as the *Mona Lisa*. Wood and other self-proclaimed "regionalist" painters insisted that midwestern artists should paint landscapes and scenes with which they were intimately familiar. Although regionalism became a nationwide sensation in the 1930s, it also provoked plenty of detractors in Iowa and elsewhere, and fierce controversy over regionalists' depiction of midwestern life foreshadowed the demise of their effort to create a distinctly midwestern culture.

Both *State Fair* and regionalist painting gained nationwide popularity in the 1930s because they offered Americans reassurance that the combined wallop of urbanization, industrialization, and the Great Depression had not destroyed the nation's 160-acre bastions of agrarian virtue. Amid the economic calamity of the 1930s, many Americans found solace in the belief that the Midwest remained a repository of the Jeffersonian creed, a land of hard-working, humble farmers untainted by urbanization and sophistication. In regionalist novels and paintings, the Midwest's fertile landscape became a product and an image to be packaged and shipped like a nerve tonic to soothe Americans' anxieties at a moment of wrenching economic, political, and cultural stress. Regionalist writers and painters sought to rekindle the Jeffersonian faith in the virtue of agriculture to counter negative stereotypes of the Midwest, and insisted that farm life remained appealing to Iowans and indispensable to Americans everywhere.[10]

From its inception until World War II, the Iowa State Fair prompted fairgoers, agriculturists, and journalists to debate the place of agriculture and entertainment at the fair and to consider the development of midwestern society and culture. In the second half of the twentieth century, Americans experienced cultural and economic transformations nearly as momentous as the process of industrialization and urbanization that had remade the nation at the turn of the century. Farmers continued to leave the countryside, and city dwellers moved to the suburbs. Today fewer than 2 percent of Americans live on farms, and most Iowans are not directly engaged in agriculture. Television, and now the Internet, have revolutionized American culture and entertainment for rural and urban Americans alike. The state fair strove to keep abreast of these changes and to prevent the fair from becoming a relic of a bygone era.

In the late twentieth century, Iowans worried anew that the state fair had become outmoded and wondered whether it would survive, just as they had a century before. In the 1980s and 1990s, some Iowans asked whether the century-old fairgrounds had become obsolete and too expensive to maintain. Some advocated relocating the fair to a new site, with new buildings and up-to-date infrastructure. But the overwhelming majority of Iowans considered the fairgrounds far more than a collection of buildings, and donors large and small contributed money to preserve them.

At the outset of the twenty-first century, the fair remains an extraordinarily popular event and institution, one that has not been consigned to obsolescence by sweeping cultural and economic change. The fair still displays Iowans' products and accomplishments, but it is no longer considered a crucial vehicle for educating its patrons or building the states' agricultural economy. It still attracts visitors by the thousands with grandstand concerts, rides, and games, but long-running debate over amusements at the fair has subsided as commercial entertainment has become almost ubiquitous in American life. In a pop culture saturated by cable TV, movies, video games, and Internet downloads, strolling through the once-controversial Midway seems downright quaint. Why all the ruckus over a few carnival games, freak shows, and burlesque acts?

A trip to the fair remains as exciting as ever, yet it's also a ritual tinged by nostalgia. Created to hasten progress in the nineteenth century, fairs now gaze backward to a bygone era in which the annual journey to the fair offered a respite from the isolation and hard work of farm life. Because most Iowans no longer live on farms, the state fair now offers an annual reminder of the central place of farming in the state's history and economy and an opportunity to visit the livestock barns and see cattle, hogs, and sheep on the hoof and up close, rather than through the window of a passing automobile or wrapped in plastic in the meat aisle of the local Hy-Vee.

Despite the extraordinary transformation of American life over the past century and a half, the Iowa State Fair remains Iowa's central cultural institution. It continues to mark time and embody the state's people, their accomplishments, and their way of life. As fairgoers amble across the fairgrounds, through the livestock barns, past the food vendors, to the Agriculture Building, the Varied Industries Building, and the Midway, they invariably size up the fair, comparing it to fairs past, pondering its novel exhibits, and enjoying once again those that remain virtually unchanged from year to year, just as Iowans have done for decades.

1 The Founders of Civilization

....

The first Iowa State Fair was held, appropriately enough, in Fairfield in October 1854. As its opening day approached, visitors flocked to Fairfield, first by dozens, then hundreds, then thousands, jamming Fairfield's usually quiet streets—the town's population scarcely topped one thousand—and filling its hotels to overflowing. Peddlers and entertainers lined the roads leading to the fairgrounds, vying with one another for fairgoers' attention and money. On opening day, a crowd estimated from seven to ten thousand people, by far the largest gathering in the history of the frontier state, congregated on and around the six-acre fairgrounds. In the summer and fall of 1854, carpenters had created the fairgrounds, enclosing it with a ten-foot-high board fence. They built a shed two hundred fifty feet long on the north side of the grounds to house the fair's many exhibits and constructed stalls and pens for livestock along the remaining sides of the enclosure. A quarter-mile track, cordoned off only by rope, was graded for the trotting races. In the center of this track stood a viewing stand for the fair's officers, judges, and speakers. The fair's officers rode and marched at the head of a parade from Fairfield to the fairgrounds, and the Iowa State Fair officially began.

George C. Dixon, a lawyer and newspaper publisher from Keokuk, delivered the fair's opening address, in which he declared the exhibition a historic event for Iowa. The fair, he proclaimed, was the very "heart" of the young state, and it would annually revitalize Iowa's farmers and send them back home to the state's "extremities" inspired to achieve the "high destiny" of agriculture. Dixon confessed that he was no farmer but began his address by proclaiming "some elemental truths": "The culture of the soil is not only the primitive calling of man, but it lies at the very bottom of the social fabric of the useful arts and social advancement." Tilling the earth distinguished civilization from savagery, and a society's progress in agriculture offered a sure index of its overall development. He invoked Thomas Jefferson's belief that independent farmers furnished the wellspring of America's republican traditions, and he quoted approvingly Daniel Webster's well-known declaration that "when tillage begins, other arts follow. The farmers therefore are the founders of civilization," as the audience on the fairgrounds nodded in agreement.[1]

Dixon's insistence on agriculture's importance belied a nagging suspicion that the status of farmers was precarious, and his hour-long address included both paeans to the dignity of farming and worries about its low prestige. His declaration that agriculture was "the primitive calling of man," which lay "at the very bottom of the social fabric," tellingly suggested that farming's status was far from exalted. Dixon lamented that other industries were speeding ahead in the nineteenth century, while farming remained comparatively static, the province of millions of unscientific "dirt farmers." He predicted that the state fair would help educate farmers and "raise up agriculture to the dignity of a useful art . . . and establish it in its merited position." Dixon's ambiguous refrain, that farmers were both America's bedrock and laggards in a fast-moving era, reverberated not only across the fairgrounds, but throughout the fair's first century, as Iowans pondered the relationship between agriculture and culture.[2]

Nineteenth-century Americans commonly took the measure of their civilization by counting its technological, economic, and cultural attainments. Many midwesterners reckoned their region's progress by tallying its booming population, crop yields, heads of livestock, railway mileage, and factory output to gauge the progress made since the frontier era.[3] Nowhere was the effort to measure and foster civilization's advance more evident than at the annual state and county fair, which existed in large part to display, assess, and celebrate the progress that midwesterners had made over the past year and since the beginning of white settlement. Fairs' exhibits offered a measure of civilization's development and a vehicle for promoting economic and technological progress. State and county fairs did much more than award blue ribbons for cattle and hogs: they embodied and sought to build an economy and society in which agricultural bounty would give rise to cities, manufacturing, commerce, and, ultimately, a distinctive midwestern culture.

An Agricultural Society

Elkanah Watson organized both America's first agricultural society, an organization created to promote improved farming techniques, and America's first county fair, in western Massachusetts in 1810. Nineteenth-century Americans proudly boasted of the contrasts between their republic and European nations, and American fairs differed indeed from British agricultural exhibitions. Watson's fair, which became the model for fairs throughout the nation, aimed to be more egalitarian than its British forerunners, which were dominated by prominent landowners and livestock breeders. The staid and exclusive British model of an agricultural society, Watson observed, was "not congenial to the genius of *our country* [emphasis in original]," and so he created a fair that invited ordinary farmers to compete for premiums to improve America's crops, livestock, and manufactures. In order to educate farmers it was necessary to lure them onto the fairgrounds. As Watson also discovered, agricultural displays alone could not entice many of his neighbors to travel to the fair. To lure farmers to attend the fair, Watson's exhibitions offered "music, dancing, and singing, intermixt with religious exercises, and measures of solidity, so as to meet the feelings of every class of the community." The fair, as he observed, was not solely an agricultural exhibition but a social occasion for isolated, hard-working farm families.[4]

Settlers transplanted Watson's "Berkshire system" of agricultural societies and fairs as they moved westward, and in no region of the country did fairs become more prominent or larger than in the Midwest. The level, fertile lands of the Mississippi Valley seemed ideally suited to agriculture, and many pioneers moved to the region to earn their living by farming. These settlers not only built farms and planted crops but also began the process of transplanting their conception of civilization to the prairie. Settlers considered agricultural fairs indispensable to their effort to build their economy and culture.

On June 1, 1833, the United States government officially opened Iowa for settlement, following the defeat of the Sac and Fox Indians led by Chief Black Hawk. Thousands of pioneers moved into the southeastern corner of the new territory and began constructing homes, farms, and communities. When Iowa became a federal territory in 1838, the second bill passed by the Territorial Legislature was "an Act to provide for the incorporation of Agricultural Societies," which permitted a group of twenty or more residents in any county to charter a corporation to foster agricultural and economic development by hosting an annual fair, awarding prizes for livestock, crops, manufactures, and other items, and by distributing information about scientific agriculture and domestic manufactures. In 1842 and 1843 the Territorial Legislature passed two more bills, spelling out agricultural societies' duties in greater detail and allocating public funds to the (as yet nonexistent) county societies in order to help them promote agriculture and domestic manufactures.[5] But Iowans hardly raced to create agricultural societies in the 1840s. Settlers in Van Buren County founded an agricultural society in 1842 and held fairs in the fall of 1842 and 1843, after which the society disbanded. No other county societies were founded until the 1850s, when economic growth and government efforts to sponsor it gained momentum.[6]

In 1853 members of the Jefferson County Agricultural Society resolved "to effect the organization of a State Agricultural Society" in Iowa and invited delegates from other county agricultural societies to meet in Fairfield in December to found the new organization. Their letter of invitation conveyed both the urgency of the task and their determination to speed the state's economic growth: "There is no free state in the union save Iowa," they wrote, "in which there is not a State Agricultural Society, organized and in successful operation. . . . Is it not time for the farmers of Iowa to be aroused to the importance of such an organization in this state? Shall we be laggards in the race of improvement? Shall the resources of other states be developed, their wealth increased and their people elevated in the scale of intellectual being, and ours stand still?"

In December, representatives of several county agricultural societies convened in Fairfield, where they founded the Iowa State Agricultural Society, dedicated to "the promotion of agriculture, horticulture, manufactures, mechanics and household arts."[7]

As a private corporation, the agricultural society had to raise its own revenues from membership dues and the proceeds of its annual state fair, and it often struggled to pay its bills. The society's officers insisted that its enormous contributions to developing Iowa's economy served the public interest and deserved public funding. Many of the society's leaders believed that it should be made a department of state government, funded entirely by the state, so that it would not be dependent on receipts from the fair. In 1857 the state legislature passed a statute granting the society an annual appropriation of $2,000 and specifying its duties: the society was now required to submit an annual report detailing its activities and assessing the condition of agriculture in the state and to collect annual reports from the growing number of county agricultural societies. County societies, in order to receive their annual appropriation of $200 from the state, were required to compile information about the condition of agriculture in their county.[8]

The state agricultural society's new responsibilities enhanced its importance and made it a clearinghouse for agricultural information in Iowa. Many Iowans mistakenly assumed that the society was a state agency, funded generously by tax revenues, and grumbled that its officers held lucrative patronage jobs. In fact, the society received only nominal aid from the legislature, and its survival depended almost entirely on receipts from the annual state fair. Some society members urged the legislature to make the organization a full-fledged department of the state government, while other hoped that it could become self-supporting and cut its ties to the state. A hybrid—part state agency, part private corporation—the Iowa State Agricultural Society alternately benefited from and was burdened by its ties to the state government throughout its history.

Lawyers, Doctors, and "Sich"

The men who led the state agricultural society considered themselves "agriculturists," who sought to promote the adoption of scientific agriculture by Iowa's farmers. But to be an agriculturist was not necessarily to be a farmer, and the society's officers were hardly a representative cross-section of the state's citizens. Its first president, Thomas W. Clagett, had served in the Maryland House of Delegates before moving west to Keokuk to practice law and publish a local newspaper. Its first secretary, Dr. Joshua M. Shaffer, a twenty-four-year-old physician who moved to Iowa from Pennsylvania in 1852 to establish his medical practice, confided to a friend that he knew "no more of farming than a hog does of fast day." Throughout its history, the society's leaders included many livestock breeders, implement dealers, large landowners, investors, and politicians. Among its officers were future US Senator George Wright, two chief justices of the Iowa Supreme Court, and numerous state legislators and other officials. Many of these men owned farms and took a keen interest in the progress of scientific agriculture, but most of them earned their living sitting behind a desk, not following behind a plow. The absence of farmers among the society's directors frequently led the state's newspapers and agricultural periodicals to charge that businessmen, not farmers, ran the agricultural society and the state fair, and that the society and the fair did not serve farmers' interests.

Secretary Joshua Shaffer replied that running an agricultural society was a task too complicated to leave to farmers. Overseeing an agricultural society, he pointed out, entailed keeping records and accounts, negotiating with local politicians and merchants, corresponding with exhibitors, and booking and promoting the fair. As a result, he stated, "Lawyers, doctors and 'sich' are uniformly called upon to do the executive work of our Agricultural Society—not because they are supposed to know anything about agriculture—but because they understand business. . . . If professional men were to withdraw their aid from our agricultural societies, as our farmers generally are & have been educated, the whole fabric would fall into ruin."[9]

Agriculturists were eager to foster the growth of agriculture, industry, and commerce in Iowa and were irrepressibly optimistic in promoting the state's image. They were unabashed economic boosters, but they also considered themselves proponents of science. "Scientific agriculture," as agriculturists understood it, mixed business acumen and scientific expertise, and agricultural societies were instrumental in introducing and fostering both commercial agriculture and agricultural science in the Midwest. Agriculturists urged farmers to operate like other businessmen by seeking ways to maximize profits by minimizing costs and labor. They also worked to develop the region's economy by instructing "dirt farmers" in "book farming" so that the Midwest could fulfill its promise as a land of bounty. Agricultural society meetings included lectures on topics such as soil acidity and livestock diseases, and the society frequently distributed publications on scientific agriculture to farmers. The state agricultural society's most important task, however, was not to host lectures or distribute scientific treatises but to run the annual state fair, which promoted agricultural improvement by bringing together farmers and rewarding exemplary livestock, crops, and products.

The state agricultural society was not the only agency created in the 1850s to foster agricultural development in Iowa. In 1858 the legislature chartered the State Agricultural College and Model Farm, located west of present-day Ames. Suel Foster, an accomplished agriculturist and member of the state agricultural society who led the effort to establish the college, considered the new institution vital for promoting scientific agriculture and arresting the growth of "citified" habits among rural youth by imbuing them with an appreciation for rural life and a strong work ethic. The agricultural society's officers championed the creation of the agricultural college, and many of them served on its board of trustees. Despite the considerable ties between the agricultural society and the college, however, the two institutions had different strategies for developing Iowa's agricultural productivity, and they soon became rivals.[10]

During its first years, the Iowa State Agricultural College and Model Farm consisted of little more than a small experimental farm. In 1862, however, Congress passed the Morrill Land Grant College Act, which allocated money to the states to create colleges to train students in agriculture and the mechanic arts. As the Iowa legislature prepared for the college to begin offering courses, it weighed the respective roles of the agricultural college and the agricultural society. The agricultural society's officers, who considered their organization the premier agency for developing Iowa's agricultural economy, drafted a proposal urging the state legislature to allow the directors of the society to "take charge of the affairs" of the college so that they "could have in charge the entire agricultural interests of the State." According to Joshua Shaffer, this plan "would bring the College Farm directly in contact with professed agriculturists." Two of the society's most eminent members, lawyer George G. Wright and agriculturist Peter Melendy, persuaded Shaffer to shelve his proposal, fearing that it would backfire and cause the legislature to diminish, rather than enhance, the agricultural society's authority. Shaffer confided to one society member that he had prudently decided to lie low, rather than tempt the "Gen 'Ass'" to alter the agricultural society's role.[11]

As Common as Courthouses: County Fairs

County agricultural societies and fairs grew rapidly in number during the second half of the nineteenth century, although these societies and fairs sometimes struggled to survive. Iowa's economic development took off in the 1850s, and the number of county agricultural societies began to grow: seven county agricultural societies were founded in 1852 and six more the following year. In 1855 Iowans held twenty-five county fairs; five years later, there were sixty-nine. As one agriculturist boasted, "Fair Grounds have become about as common as Court Houses."[12] But competition from nearby fairs, poor weather, economic downturns, the Civil War, and dissatisfaction with the fairs' management drove many fairs and agricultural societies into debt or even bankruptcy. The Des Moines County society, for example, reported in 1858 that it was "sometimes in existence and sometimes not." Its counterpart in Dubuque County confessed that it "had a nominal existence in this county for some time but has never amounted to anything." The task of compiling an extensive report on the condition of agriculture in their county proved an enormous burden for many of these small, poorly funded county agricultural societies. Joshua Shaffer complained in 1869 that too many county reports were incomplete, inaccurate, and prepared hastily only "to secure an appropriation from the State, [rather] than to give a synopsis of the agricultural condition" of the county. To his annoyance, many county reports described in detail the annual fair but neglected to provide statistics on the county's crops and livestock.[13]

In an effort to create larger, more successful fairs, some county agricultural societies banded together to form district agricultural societies, which joined agriculturists and farmers from several counties into a single organization. As one district fair secretary pointed out, many counties simply could not mount a successful exhibition, and creating district societies enabled agriculturists to join forces to host a bigger and better fair. Another district fair secretary wrote in 1880 that "there are too many fairs and too few creditable exhibitions. If two-thirds of the county societies would disband and re-organize on the district plan, embracing a larger population in such concentrated effort, the fairs would be more popular and useful." In addition to the county societies formed in each of the state's ninety-nine counties, more than eighty district societies were created in the second half of the nineteenth century. Some of these district societies went broke in a year or two, while others thrived for decades. Some linked only two or three counties, but the largest district societies, and their fairs, were enormous. The Cedar Valley District, under the direction of Peter Melendy of Cedar Rapids, the state's most eminent agriculturist, included eleven counties, while the Northwestern Iowa District, which held its first fair in Fort Dodge in 1873, comprised thirteen. The Central Iowa District, formed in 1860, grew until by 1870 it encompassed nineteen counties, which were home to one-quarter of the state's residents.[14]

District fairs attracted thousands of fairgoers, and district fair secretaries often claimed that their exhibitions rivaled, or even surpassed, the state fair. This was no idle boast: some of these larger fairs posed a serious challenge to the fledgling state agricultural society, which declared that district societies were "not desirable," ostensibly because they undercut farmers' allegiance to their county societies. But by the 1880s the state agricultural society and state fair had grown large enough that district fairs no longer threatened the state fair's preeminence or receipts. District societies now seemed preferable to attempting to host a successful fair in each of the state's ninety-nine counties. "The day is not far distant," wrote the state agricultural society's secretary, John Shaffer, nephew of Joshua, in 1884, "when the number of our county fairs will have to be reduced and organized into districts. Too many now live on the state's appropriations, and even with this too many are failures." The number of county and district agricultural societies continued to grow throughout the nineteenth century, and agriculturists proudly tallied their number, just as they tallied increasing crop yields, as evidence of the state's rapid progress. In 1880 ninety-nine county and district societies filed annual reports with the state agricultural society; by 1893 one hundred fifteen societies served the state's ninety-nine counties. Fairgrounds had become even more common than courthouses, and fairs were a staple of Iowa's culture.[15]

County fairs, like the state fair, were considered vehicles for improving agricultural productivity and barometers for measuring a county's attainments in agriculture and manufactures. One fair secretary saluted the county fair as "a camp ground annually tented in the defense and promotion of the great cause of agriculture, mechanism, and every art that tends to advance the interests of the farmer, and elevate and secure to him the true and noble position that nature has assigned him." According to Joshua Shaffer, as secretary of both the Jefferson County Agricultural Society and the state agricultural society, a county fair "should be the representative of the agricultural condition, resources, capacity of improvements, etc., of the county."[16]

Agriculturists publicly extolled the virtues of farming, but privately they grumbled that too many farmers were reluctant to participate in their local agricultural societies, exhibit their products at the fair, or adopt new ideas about scientific agriculture. The secretary of the Jones County Agricultural Society, Dr. J. S. Dimmitt, complained that "attendance at the first Jones County Fair was *small*, citizens' interest in the exhibits *smaller*, and the prospects for building up an Agricultural Society, the *smallest* [emphasis in the original]." Some farmers were not merely apathetic but downright resistant to scientific advance. Isaac Kneeland, secretary of the Lucas County Agricultural Society, complained that "our farmers are poor agriculturists, with few exceptions. They are not reading men, and are generally disposed to plow as shallow as they can." Similarly, E. T. Cole, secretary of the Davis County Agricultural Society, declared in 1866 that too many farmers "are content to plod along in the same dull routine of work that their fathers and neighbors did." To agriculturists' dismay, some farmers posed an impediment to developing the state's economy.[17]

Although county fairs were created to display and foster agricultural and economic progress, these exhibitions sometimes indicated that the task of transplanting "civilization" and scientific farming to the frontier had barely taken root. One Montgomery County resident later recalled that the county's first fair resembled a band of "shipwrecked and homeless people on a barren island," rather than an oasis of civilization on the prairie. Farmers were sometimes reluctant to haul exhibits to the fair only to be subjected to what one agriculturist called "the scathing ordeal of public opinion." As a result of farmers' unwillingness to exhibit their products, the paltry exhibits at the first Marion County Fair in 1853 comprised only a few cattle, a dozen horses, a tiny display of field crops, and a pair of quilts. Visitors surveying the display of housewares at the inaugural Montgomery County Fair saw only "a cheese, some butter and a few tin cans containing sorghum syrup." Fairgoers could hardly be expected to glean much knowledge or inspiration from exhibits this meager.[18]

County fairs were predicated on the assumption that farmers learned most effectively about scientific agriculture by looking at animals, objects, and exhibits and then emulating their neighbors' examples. The goal of a fair's displays, as one agriculturist put it, is "not to find out which is the best horse, cow, or other exhibit, but to educate ourselves so we may be able to appreciate the good qualities and detect the bad ones." When a farmer walked through the fair's livestock pens, he would almost invariably consider what qualities distinguished a blue-ribbon cow from the herd. Judges for the exhibits at county fairs were usually appointed by the "pick-up system," in which local farmers and local residents were enlisted to judge animals and exhibits. Reliance upon amateur judges expressed faith in farmers' ability to assess the merits of animals, crops, or handicrafts. Because the judges themselves were ordinary farmers, "the onlookers feel free to discuss the merits of the awards, which is the best school." Encouraging fairgoers to form their own opinion of the worthiness of the fair's exhibits sometimes generated controversy over the ribbons awarded in the fairs' competitions. But county fairs continued for decades to adhere to the premise that local farmers and resident were competent to judge the fair's exhibits and awarded prizes to animals and exhibits that embodied the standards and tastes of their community.[19]

The State Fair

The Iowa State Agricultural Society promoted economic growth, lobbied state legislators, sponsored addresses and publications on scientific agriculture, and published a hefty annual report. But it was known to Iowans almost exclusively for one thing: running the state fair. The fair remained the society's main vehicle for reaching the state's residents and disseminating its vision of scientific agriculture, and on the fair's success rested the society's finances and reputation.

The fair was not always the ideal venue for promoting scientific agriculture and economic development. Despite agriculturists' lofty pronouncements about the fair's educational mission, the fair mingled agricultural and commercial exhibits with sideshows and festivity. While the fair's exhibits of crops and livestock celebrated agriculture, labor, and productivity, many of its entertainments embodied festivity and even outright unruliness, and represented an economy devoted to leisure and consumption. The fair enticed farmers, their families, and other patrons to browse and to enjoy themselves, not to study, and it attracted scores of merchants and showmen who took little or no interest in scientific agriculture. Part scientific exhibition, part bazaar, part carnival, the fair's boisterous fun ran counter to the society's effort to host a scientific exhibition promoting its vision of an orderly and prosperous agricultural economy.

Billing the exhibition as a state fair did not make it one in fact, and the state agricultural society strove for years to make the state fair worthy of the name and to win Iowans' allegiance. Organizing, publicizing, and operating the fair posed an enormous task for the society's officers, and the fledgling agricultural society confronted a host of difficulties in attracting exhibitors, enlisting judges, publicizing the fair, and paying its bills. The state fair often had to compete with county and district fairs and occasionally with neighboring state fairs. Although the fair was designed to promote a vision of economic stability, it was a fleeting event. As John Shaffer, who worked countless hours to organize the fair each year, said resignedly in 1876, "The fair is utterly evanescent in its character. Thousands of dollars are expended to prepare for it. It is arranged, organized, completed, and dissipated in a week. Then the halls and stalls and buildings are left to decay and weeds and neglect for another year, when the same process is repeated."[20]

During the fair's first two decades, poor transportation and inadequate facilities on the fairgrounds made many farmers reluctant to transport their livestock, crops, and other items to the fair. As a result, the "state" fair remained primarily a local event for several years. "With the exception of horses and a few sheep," the *Iowa Farmer* lamented after the 1855 fair, "we heard of no livestock exhibited that came beyond twenty-five miles, and but few animals much over half that distance. The agricultural productions were mostly confined to Jefferson and the counties immediately contiguous." The fair soon began to attract livestock breeders and implement manufacturers from across the Midwest, who hoped to profit from showing their animals or selling their machines at the fair, but the fair's "lesser" divisions remained local competitions for decades. As Joshua Shaffer wrote in 1869, "the exhibition in the minor classes is necessarily local in its character. In pantry stores, domestic manufactures, mechanic arts &c &c the display is mostly made up by persons convenient to the Fair. In the heavy classes, of Stock, Implements, Machinery &c a dozen states may be represented. Dealers and stock-men will not miss the opportunity to exhibit."[21] Attendance, too, was confined largely to those who lived near the fair. In 1871 Keokuk's *Daily Gate City* bluntly declared that no sensible person would travel more than thirty or forty miles to see the fair, concluding that "there won't be a crowd unless the Fair is in a populous community and country."[22]

Because people and exhibits would not necessarily come to the fair, the fair came to them. The agricultural society moved the exhibition to a new town every second or third year in order to afford a larger number of the state's residents at least an occasional opportunity to attend the fair and to satisfy local businessmen, who profited immensely when their town hosted it. Competition between cities vying to host the fair often became acrimonious at the society's annual meeting, even though its officers had usually struck an agreement about the fair's location months before the society's convention. During its first twenty-five years, the Iowa State Fair was held in ten cities:

Fairfield	1854–1855
Muscatine	1856–1857
Oskaloosa	1858–1859

Iowa City	1860–1861
Dubuque	1862–1863
Burlington	1864–1866
Clinton	1867–1868
Keokuk	1869–1870
Cedar Rapids	1871–1873
Keokuk	1874–1875
Cedar Rapids	1876–1878
Des Moines	1879–

Moving to a new fairgrounds every two or three years entailed enormous effort and expense, and the society repeatedly urged the state government and local boosters to shoulder the cost of equipping the grounds. Secretary John Wallace, who oversaw the fair from 1856 to 1862, worried that local businessmen and governments would refuse to contribute to building the fairgrounds, and that the society "would soon go begging for places to hold her exhibitions." But, as the fair's attendance and value to local businesses grew, the society was able to exact concessions from the towns bidding to host the fair.[23]

As settlers moved into the state's western and northern counties, many wanted the fair to move with them. In 1870 some Iowans grumbled that Keokuk, located in the state's southeastern corner, offered a singularly poor site for the state fair, and the *Iowa State Register*, published in Des Moines, derided the exhibition as "the Illinois-Missouri-Iowa Fair." After the fair closed, the state's leading agricultural magazine, the *Iowa Homestead and Western Farm Journal*, also published in Des Moines, wrote that "in several respects it was not a State Fair, either in the animals or the articles exhibited, or those who attended the exhibition. . . . But we must hereafter have it an Iowa Fair, not only in name, but in fact, if it is intended that the industrial interests of the State shall be encouraged." The magazine claimed that Iowans had grown dissatisfied with the "Illinois Fairs" held along the state's eastern border and hailed the society's decision to move the 1871 exhibition westward to Cedar Rapids. "As our prairies become filled up with tillers of the soil," the magazine predicted, "these exhibitions will be removed farther west, to shed their beneficent influences on others who may need them."[24]

Whatever its geographic disadvantages, Keokuk outmaneuvered Cedar Rapids, Dubuque, and Burlington to regain the fair in 1874. The *Cedar Rapids Times* remarked bitterly that Keokuk was "the most out of the way place in all Iowa, the farthest removed from the rest of the State, being situated in a notch cut into Missouri, and a far more appropriate place for holding a fair for that State than for Iowa. After wandering around the State, dodging into out-of-the-way holes and corners like Keokuk, we will welcome it back to the best location in all Iowa, and again place it in a condition which will be a credit to the State."[25] The editorialist was right on several counts: Secretary John Shaffer acknowledged in the summer of 1874 that he had received more requests for the fair's premium list from residents of Illinois and Missouri than from Iowans. When the society's treasurer tallied the expenditures and receipts from the 1875 fair in Keokuk, the organization was nearly broke. The *Cedar Rapids Times* reiterated its litany of the merits of Cedar Rapids and drawbacks of Keokuk, inviting its readers to draw the obvious conclusion, and from 1876 to 1878 the fair did return to Cedar Rapids. But boosters in Des Moines eagerly sought to bring the fair to the state's capital city. Centrally located and well served by railroads, the city offered an advantageous location for the fair, which was moved to T. E. Brown's Park, a "driving park"—that is, one equipped with a track for horse races—in 1879. Although other cities vied to host the fair over the next few years, it remained in the capital city.[26]

System in the Exhibition

Agriculturists maintained that the fair was primarily a scientific exhibition designed to foster the adoption of better agricultural techniques, purebred livestock, and farm implements. Because agriculturists considered themselves proponents of science, early fairs offered premiums not only for animals, fruits, and vegetables but also for the best treatises on agricultural science, and their schedule included addresses on soil chemistry, livestock diseases, and horticulture. But the *Iowa Farmer* doubtless spoke for many farmers when it suggested that the fair could accomplish more good by omitting ponderous lectures by agriculturists and rewarding instead the tangible results of farmers' labors. Distrustful of "book-farming," the magazine advised agriculturists to promote practical knowledge by rewarding only actual products at the fair. An exhibit of animals, vegetables, or grains, the magazine declared, "is far purer and better than the hired labors of the pen unconnected with the plow—keep the pen and the plow close together, for there are quacks in these days."[27] State and country fairs were predicated on the assumption that most people learned best by seeing exhibits of livestock, crops, or crafts that would inspire them to emulate the most outstanding examples. As the society's president, George G. Wright, explained in his opening address at the 1860 fair, the fair's exhibits served

> to *embody into practical, material form, the ideas* which would *else exist only in the minds as theories*; that as practical men and women, we may derive lasting and solid advantage and instruction. Mere abstract truths and theories were of little use in the world until they were wrought out into practical result, and a Fair should, therefore, be a practical workshop; a place where facts were materialized so that they could be seen and felt and handled [emphasis in the original].[28]

One of the society's subsequent presidents put it more succinctly: "Word pictures may be appreciated by many, but the thing itself in actual view is the great instructor to the masses of the people."[29]

In its annual premium list, the society spelled out in detail the items worthy of inclusion at the exhibition. These lists, organized into categories ranging from livestock and machinery to handicrafts and oil paintings, furnished a blueprint of the society and culture that agriculturists sought to build. The premium list of the first fair in 1854 offered prizes of $1,100 for more than four hundred different items.[30] Initially, most of the fair's winners did not receive a cash premium but were rewarded with the society's diploma, which the society deemed its "highest honorary award." The 1857 premium list explained that the society's officers "have decided that it is not advisable to pay any premiums in money, except where the circumstances of the case render it indispensable." Appeals to farmers' sense of pride, unfortunately, failed to attract many exhibitors to the fair, and as its finances permitted, the society increasingly rewarded exhibitors with cash premiums. In an address to the society in 1864, President George G. Wright acknowledged that "It is easy to *talk* about the *motives* that should and ought properly to influence exhibitors in coming to Fairs . . . and yet, practically, we well know that without *money* we cannot succeed. . . . The more money and the larger the premiums, the greater the competition and the more gratifying the success [emphasis in the original]." Without generous premiums to entice exhibitors, he warned, "the gates of our fair grounds would rust upon their hinges and our halls decay from non-use."[31]

The premium list also spelled out the fair's schedule and rules, which were enforced by a host of judges and other fair officials. But the fair's myriad activities and competitions and the throngs of fairgoers milling about the grounds often impeded the society's effort to conduct the fair in an orderly and punctual manner. The first two fairs proved so haphazard that the society's directors despaired "that it was impossible to bring the people of Iowa to anything like time or system in an agricultural exhibition." In 1856 the society resolved to conduct a more orderly fair, appointing a chief marshall to ride herd over the fair's exhibits and ensure that its rules and schedules were carried out "to the letter in every particular." The results were encouraging: "SYSTEM IN THE EXHIBITION has been demonstrated to be practicable. . . . In no instance did the exhibition of an animal or article vary more than five minutes from the time fixed upon for its exhibition." The society boasted that it had rendered the fair nearly as orderly as the premium list itself [emphasis in the original].[32]

The Choice of Judges

The most important work at the fair fell to the judges of the various competitions, who determined which animals, produce, and other items deserved to be awarded premiums. When the state agricultural society was founded, the *Iowa Farmer* emphatically advised the new organization that appointing capable, honest judges, especially for the fair's livestock contests, was indispensable in order to earn farmers' respect. Secretary John Wallace noted in 1858 that "the usefulness, if not the very existence of the society, depended upon the choice of Judges." In his opening address at the fair, the society's president typically urged exhibitors to abide by the judges' decisions, however much they might disagree with them. Inevitably, however, competitors sometimes disagreed with the judges' evaluation of their animals, crops, or products, and exhibitors lodged protests against the judges' decisions almost the moment the fair awarded its first premiums in 1854. Irked by dozens of complaints of judges' incompetence and favoritism at the inaugural fair, President Thomas W. Clagett harshly rebuked several "mean and contemptible" contestants for daring to question the judges' integrity, reminding them that the judges "are human beings, like yourselves, and are liable to make mistakes and accidental omissions."[33]

The fair's judges were not always beyond reproach, and the society sometimes appointed judges who were unfamiliar with the animals or items they were assigned to evaluate. In order to identify potential judges for the fair's exhibits, the secretary routinely asked the fair's exhibitors to suggest names of "a few of your best farmers, stock growers,

&c." To compound the society's difficulties, judges sometimes failed to attend the fair, forcing the society to recruit replacements on the spot. When Secretary John Wallace called the roll of judges at the 1860 fair "for the purpose of filling such vacancies as might appear," barely a third of the judges, even those for the fair's most important livestock competitions, were present, compelling him to call for volunteers from the crowd. As the *Iowa City Republican* glumly reported, "In the haste of such a moment, grave mistakes are often committed, men being placed upon Committees who are utterly unfitted for the particular duty assigned to them. Hence follow unjust awards, and dissatisfied exhibitors."[34] After the 1868 fair, one judge confessed that he had awarded premiums for "a branch of stock that he did not know anything about." He claimed that listening to "the wrangling of exhibitors" had helped him select the eventual winners but admitted that "it was rather a rough joke on the State Fair, and the exhibitors too."[35]

The fair's livestock exhibitors were knowledgeable and opinionated about their animals, and they stood to gain premiums, prestige, and profits from winning the fair's competition. They had little patience with inept judges, who sometimes rewarded inferior animals. After the 1873 fair, the *Iowa Homestead and Western Farm Journal* blasted the incompetence of many of the fair's livestock judges, fuming that

> breeders are made to smart under the rod held over their heads by committee men who are not in any sense competent to do justice to so important an interest. Many an animal goes home with a commercial value much less than when it enters the grounds, that is, allowing the public to be the judge, because beaten in the ring by a beast not representing a fourth of its value. And in many cases, an unmerited value is heaped upon a miserable, unworthy thing . . . enabling the owner to sell third-rate stock for first-class price.

Breeding and raising livestock was a lucrative business, and stockmen competed at the fair in order to boost their profits, not just to collect ribbons.[36]

Since the days of Cain and Abel, raising livestock had been considered more prestigious than growing crops, and exhibits of livestock, especially beef cattle, had long been the mainstay of European and American fairs. Agriculturists considered livestock judging the most important of the fair's activities, and the competitions for cattle, horses, hogs, and sheep headed the fair's premium list. Agriculturists believed that raising purebred livestock and eradicating "scrubs" was essential to developing the Midwest's agricultural economy, and so only purebred livestock were allowed to be exhibited at the fair.[37] Although corn quickly became Iowa's staple crop, most corn had not yet been harvested by fair week. (The gap between the fair and harvest season grew wider over time as the dates of the fair were moved earlier. The fair was initially a harvest festival, held in October, but within a few years it was moved to September. Eventually it was scheduled in August to accommodate summer vacationers.) After the 1868 fair, Joshua Shaffer lamented that, although corn was the state's principal crop, "there has seldom been an exhibition in this regard that would do credit to a third-rate county fair." Exhibits of fruits and vegetables were important at the fair but did not rival the prominence of the livestock contests.[38]

The fair's livestock contests provoked controversy over the criteria that determined which animals deserved the blue ribbon, and whether animals should be judged according to the standard of farmers, agriculturists, breeders, or animal scientists from the State Agricultural College. In the fair's first two decades, farmers and agriculturists generally agreed that superior cattle and hogs were the heaviest and densest, regardless of their other merits or defects. An early premium list informed exhibitors that "all other things being equal, those are the best cattle that have the greatest weight in the smallest superficies." Many farmers and judges simplified the society's formula: the best cattle were those that were just plain biggest. Farmers and judges commonly equated excellence with sheer bulk, even though no prudent farmer would fatten his cattle or hogs to the freakish size of show animals, which often resembled four-legged dirigibles.

Farmers' and agriculturists' standard for judging livestock encountered significant challengers in the 1870s and 1880s. Professors of agricultural science at the State Agricultural College scorned farmers' and amateur judges' preference for fat cows and pigs and insisted that excessively fat stock were unhealthy, undesirable, and, worst of all, unprofitable. Beginning in the 1870s, college faculty members urged agriculturists to accept their new scientific standards of assessing livestock. In 1873, college president A. S. Welch implored the agricultural society not to allow ill-informed amateur judges to award prizes to excessively fat animals and to employ only expert judges, either university-trained experts or professional stockmen, to oversee the fair's livestock competitions. Welch urged the society to resolve "that no special preparation for the annual show, no ornamentation, no bedizenment, no superfluous fat, should ever blind the eyes of the judges, and hide the lack of genuine merit." He believed that state and county fairs could contribute to agricultural progress but complained that they had too often "given encouragement, unwittingly, to a line of policy which has divorced the fair from the market, and led the producers to prepare, by different processes, one thing to sell and another to show." In Welch's view, the fair ought to reward only those animals that were raised to supply the demands of the market, and he proposed that the fair's judges calculate the difference between the amount of money spent to raise an animal and the amount that a farmer would receive for selling it, when awarding premiums.[39]

From his vantage as head of the agricultural college, Welch eyed the state fair skeptically as a Barnumesque "museum of mere curiosities" rather than a legitimate scientific exhibition, and he sought to educate the self-styled scientific agriculturists who ran the fair. Welch addressed the crowd at the 1874 fair in Keokuk (between horse races at the fair's track!), repeating his charge that the fair's judges too often awarded prizes to "coarse cattle, covered by years of pampering with layers and tumors of fat," so that "no Christian could eat their flesh." Bestowing premiums on such outlandish animals degraded the fair, turning it into a "menagerie," rather than a scientific exhibition. Welch reiterated his case for employing expert judges, trained to assess animals, crops, and other items at the fair according to a "safe and solid standard." For Welch, the only reliable standard for measuring the worth of livestock and crops was "that delicate test of value, the market price." Many of the society's officers shared Welch's belief that agricultural production should be tailored to meet the demands of the market, but they also recognized that the college's growing authority as a scientific institution posed a challenge to the agricultural society's role in fostering Iowa's agricultural and economic development.[40]

Complaints about the fair's judges were not merely a nuisance for the agricultural society but called into question the fair's usefulness in improving the state's livestock, crops, and other products. Disgruntled livestock exhibitors seldom confined themselves to polite disagreement over the finer points of their animals. One Clydesdale breeder, incensed by the judges' low estimation of his horses at the 1883 fair, hurled his ribbons at the superintendent of the fair's horse department while spluttering curses at the superintendent of the fair's horse exhibits, the judges, and the society's board of directors.[41] Even more worrisome than dissatisfied individual exhibitors were the protests lodged by breeders' associations, which promoted particular stock breeds, and whose approval and participation were indispensable to the fair's livestock competitions. The American Berkshire Association, for example, declared after the 1882 fair that its members "are little disposed to place their animals in competition where no account is taken of the superior worth for breeding purposes of well bred stock." Livestock breeding had become an enormous and lucrative business, and breeders understandably resented the fair's unsystematic method of judging their animals. Livestock exhibitors and breeders, frequently annoyed by the fair's amateurish and unscientific method of awarding premiums, heartily endorsed Welch's recommendation that the society hire expert judges.[42]

Despite the fair's ostensible commitment to ensuring that ordinary farmers had a reasonable prospect of competing in the fair's contests, the biggest premiums in the fair's livestock divisions were increasingly carried off by professional breeders, who lavished extraordinary attention on their magnificent animals and traveled from fair to fair to compete for premiums. In an effort to ensure that prizes in the fair's livestock and machinery competitions would not be won exclusively by professional livestock breeders and implement manufacturers, state law required agricultural societies to host fairs that would permit "small as well as large farmers and artisans to compete." Since Elkanah Watson's day, American agricultural fairs had sought to include typical farmers in their competitions, and to avoid becoming a showcase solely for gentleman farmers and professional livestock breeders. Initially, the state fair's livestock contests were restricted to Iowa farmers, but by the early 1860s the society had declared its contests "open to the world."[43]

According to one journalist, professional breeders soon became so dominant in the fair's competitions that "the ordinary stock raiser has no more chance of winning a prize on one of his animals in the show ring against these pampered and often barren pets than as though he showed a rabbit." Small farmers often resented large livestock breeders, who dominated the fair's competitions by exhibiting their immaculately groomed, almost unnatural specimens. The freakish "pampered and barren pets" exhibited by professional stockmen were raised so differently from most livestock that the fair's competitions seemed divorced from the workaday world of the typical farmer. Most farmers were obliged to raise their herds efficiently in order to earn a profit, while breeders lavished attention on a single "show animal."[44]

By the 1880s dissatisfaction with the fair's judges, coupled with pressure from the agricultural college, impelled the society to begin hiring professional judges in its livestock competitions. The fair employed professional judges in its poultry competitions as early as 1874, but the livestock contests remained in the hands of committees of three judges, who, whenever possible, were expected to have some familiarity with the breeds they were appointed to assess.[45] In 1884 the society began hiring experts from outside the state (because non-Iowans were less likely to be accused of partiality) to judge its beef and dairy cattle contests. Impressed by the results and by exhibitors' willingness to abide by the judges' decisions, Secretary John Shaffer decided that expert judges should be appointed to oversee *all* of the fair's competitions. Shaffer became so enthusiastic about the potential of expert judges that he declared that livestock judging could be made into an exact science. Delighted by the professional judges' performance at the 1884 fair, Shaffer predicted that "the day is not distant (especially in stock) when an expert will be demanded for each particular breed," and he suggested that the fair create "report cards" on which judges could objectively score the merits of each animal. These report cards would prove instructive to livestock exhibitors, make excellent advertisements for their animals, and "add renown to our Society that we 'Judge by Scale.'" In 1885 the society hired experts to oversee the fair's horse, swine, and sheep judging and created scorecards for these contests. But Shaffer soon acknowledged that some aspects of livestock judging remained unquantifiable and perhaps even subjective. Animals, after all, were "shown" in a ring, and judges and exhibitors might reasonably disagree over the qualities that distinguished a blue ribbon animal from a runner-up. Animal judging was a job for experts, but it was hardly an exact science.[46]

Employing professional livestock judges preserved the credibility of the fair's livestock exhibitions but also attested to the fact that college-trained agricultural scientists had eclipsed the fair's self-proclaimed agriculturists as leaders of the state's agricultural development. The agricultural society, which had formerly deemed its own members and ordinary farmers competent to judge animals, crops and other exhibits, now deferred to the authority of scientists and experts. The fair's original premise, that farmers could educate themselves by forming their own estimation of animals and exhibits, yielded to a new conception of scientific agriculture, overseen by university-trained scientists.[47]

Manufacturing Interests

The Iowa State Agricultural Society sought to improve Iowa's crops and livestock, but its vision of a vibrant economy entailed more than developing the state's agricultural productivity. The society's officers sought to create an economically diversified, self-sufficient state, one that contained factories as well as farms, and they considered increasing Iowa's agricultural output only the first step toward creating a diversified and prosperous economy. While the society's officers frequently extolled the virtues of agriculture, they were more Hamiltonian than Jeffersonian, and they considered manufacturing and commerce indispensable to Iowa's prosperity. "An exclusively agricultural state can never be rich and prosperous," wrote Joshua Shaffer in 1866. "There must be a diversity of labor, and a multiplication of channels for the investment of capital."[48] Shaffer believed that "the manufacturing interests of our State should have large inducements offered to exhibit their goods at the State Fairs," and the society's premium list offered prizes not only for livestock, grains, and produce but also for manufactured goods ranging from eyeglasses to bricks to farm implements.

Developing Iowa's manufacturing proved at least as difficult as increasing its agricultural productivity. After the displays of "mechanics' goods" at the first two state fairs proved disappointing, the society urged the state's inventors and craftsmen "to carry something with you" to the fair, "to swell the number of contestants, and to fill the grounds with the product of your factories, your shops, and your workhouses." The following year, Secretary John Wallace noted optimistically that mechanics and manufacturers were "beginning to realize the value of this method of advertising their wares" and predicted that the upcoming fair would bring them out "in all their strength."[49]

While the society's officers were eager to see Iowa develop a wide variety of industries, they were especially eager to promote the manufacture of farm implements. Because Iowa's economy seemed destined to remain primarily agricultural, the production of farm machinery offered the most likely prospect for developing indigenous manufacturing. Implement manufacturers would not only help increase the state's agricultural productivity but would keep wealth within the state by enabling farmers to purchase locally made machinery instead of "importing" plows and harvesters from manufacturers in Chicago and other cities. Ultimately, however, industrialization did not foster economic self-sufficiency in Iowa but enmeshed Iowans more tightly into the national economy, in which the major industrial centers lay to the east. As Olivier Zunz observes, implement manufacturers, railroads companies, and salesmen integrated farmers into a national market for manufactured goods.[50]

The state fair afforded farmers an opportunity to compare and purchase implements, and its machinery display immediately became indispensable to the fair's success. The *Iowa Farmer* observed in 1857 that the machinery exhibit attracts "the first attention of the real farmers, and those desiring to improve." According to the *Dubuque Democratic Herald*, "any display of the principles of the mechanic arts always has a fascination about it that attracts even the dullest minds." Implement manufacturers' exhibits not only attracted attention on the fairgrounds, but received far more coverage in the state's newspapers and agricultural journals than any other aspect of the fair.[51]

Although agriculturists were eager to promote the mechanization of farming and the development of manufacturing, they conceded that some of the contraptions displayed at the fair were gimmicks and even outright frauds. Prior to the 1867 fair, Joshua Shaffer confided to one manufacturer that the invention of farm implements had given rise to "more humbug, mixed with a few grains of gold, than any other subject." He added that, in his opinion, some of the implements awarded premiums at recent fairs were utter junk and stated that most machines did not increase farmers' productivity enough to offset their expense. Shaffer urged judges in the fair's machinery exhibit to "to discriminate between an article merely gotten up for 'show' & one for service." As with the animals in the fair's livestock contests, appearances could be deceiving, and machinery salesmen were not above using showmanship, hype, and even fraud to sell their products.[52]

Despite Shaffer's private misgivings about the utility of many farm implements, the agricultural society eagerly accommodated its largest commercial exhibitors. Because implement manufacturers were essential to the fair's success, they were treated far better than the other salesmen who hawked their wares on or near the fairgrounds. The society offered large manufacturers, free of charge, some of the most desirable space on the fairgrounds. In its annual report of 1870, the society published more than fifty pages of illustrations and descriptions of implements exhibited at the fair. Ostensibly, these illustrations were included to inform farmers about innovations in farm machinery. But several newspapers accused the society of blatantly advertising farm machinery at taxpayers' expense. "You should see how sundry newspapers are hounding me," Shaffer wrote to one manufacturer, "for *'advertising'* as they call it, the implements we had on exhibition!"[53]

In the 1850s and 1860s, farm implements were included in the fair's premium list, and judges awarded prizes to superior machines. But machinery displays differed from the fair's other contests because implement manufacturers were far more interested in making sales than winning premiums. Farmers did not amble about the machinery exhibit to satisfy their innate curiosity about mechanical devices but because the fair afforded them an excellent opportunity to compare and buy farm implements. In the 1870s, the society's officers realized that implement manufacturers benefited enormously from displaying and advertising their machines at the fair, and that no premiums whatsoever were necessary to attract them. (Tellingly, though, farm implements continued to be included in the premium list even after premiums were no longer offered.) As Secretary John Shaffer observed in 1878, "In former years large money premiums were awarded for agricultural implements and machinery; the money was taken away and the diploma offered; the result was an increase regularly every year. The diploma, too, has been taken away recently, and the exhibits are larger and finer than ever in these departments." Implement manufacturers were not enticed primarily by the prospect of winning premiums but by the opportunity to advertise and sell their machines at the fair, and the display of farm implements was no longer a competition but a commercial exhibit.[54]

Even though the society's officers understood that implement manufacturers and dealers profited from showing and selling their machines at the fair, they remained solicitous of manufacturers. When the society stopped offering premiums to implement manufacturers in 1876, it saluted the altruism of the "public-spirited gentlemen who show plows, farmers' and mechanics' tools, farm implements, farm machinery, mills, engines, &c., in classes for which no premiums are offered, and which showing is made at great expense and trouble to such exhibitors." Agriculturists considered implement manufacturers crucial to launching industrialization in Iowa and building a diversified economy atop the state's agricultural productivity. Secretary John Shaffer summed up agriculturists' estimation of the importance of the fair's machinery exhibits on the eve of the 1880 fair: "I would sooner see any other department fail than that one."[55]

The Crumbling Farmers

Agriculturists sought to promote a diversified economy in Iowa in which agricultural prosperity would ultimately contribute to the growth of commerce and manufacturing. But the interests of farmers and corporations were not always harmonious, and agriculturists' commitment to the interests of manufacturers and railroad corporations became evident when the economy declined and the Grange rose in the 1870s. The depression of 1873 spurred the state legislature to enact drastic cost-cutting measures, and it revoked the society's annual appropriation, dealing a severe blow to both the society's treasury and its prestige. The depression also prompted some blunt criticism of the agricultural society and its vision of the state's economy. The rapid growth of the Patrons of Husbandry, commonly known as the Grange, challenged the society's claim to represent the state's farmers, ran counter to its booster ethos, and cost the society some setbacks in the legislature as well.

The Grange was founded in 1867 by Oliver Hudson Kelley, a Minnesota farmer and journalist, to serve as a social and educational society dedicated to improving rural life. The organization soon became engaged in politics, advocating regulation of railroad freight rates, which were crucial to farmers' ability to market their crops profitably. The Grange rapidly attained a popularity in Iowa unmatched in other states, claiming 100,000 members (Iowa had 116,000 farmers according to the 1870 US Census!) organized into nearly 2,000 local chapters. In only a few years, the Grange had garnered the allegiance of an overwhelming majority of the state's farmers, and the organization posed a serious challenge to the agricultural society's status as the representative of the state's farmers.

Many agriculturists fretted that Grangers would gain control of the State Agricultural Society and use it to promote the Grange's political agenda. While some of the agricultural society's leaders joined the Grange, they recoiled from the Grangers' effort to regulate railroad corporations, which they considered indispensable to the state's prosperity. The Grange's political views ran counter to the agricultural society's boosterism. In the language of the day, Grangers were not boosters but "kickers." Many of the agricultural society's officers feared that the Grangers' hostility to railroad corporations would discourage investment and economic growth by making the Midwest unattractive to railroads, manufacturers, and investors. As Joshua Shaffer observed,

"Capital likes quietness. Noise & commotion are death to all money ventures." His nephew John privately blamed "the grumbling farmers" for misguidedly supporting "Granger laws" regulating railroad freight rates. "It seems to be hereditary with them to complain," he wrote. "So must it be and it will be from now to eternity!" The Shaffers knew better than to criticize farmers or the Grange publicly, because, as Joshua Shaffer conceded, "It would be folly for me to attempt to stem this torrent that is sweeping over the State. It would drown me, and do no good." Their private correspondence, though, attests to their belief that Iowa's economic growth ultimately depended more on building railroads and attracting manufacturers than on the interests of individual farmers.[56]

A Fair Ground

Despite its varied efforts to promote Iowa's agricultural and economic development, by far the state agricultural society's most important function was to hold the annual state fair. The lack of a permanent fairgrounds impeded the society's effort to create an exhibition that would embody Iowa's prosperity, stability, and progress. Wearied by the society's chronic financial difficulties and precarious status, its officers resolved in the 1880s to make the fair more profitable, to purchase and construct a permanent fairgrounds, to enhance the society's reputation as a scientific organization, and to gain recognition as a full-fledged public institution. By the 1880s many of the society's officers had realized that the state fair would never be successful or reputable until the society acquired a permanent fairgrounds and constructed substantial exhibition buildings and other facilities for the fair. As one of the society's directors argued in 1883, shuttling the fair from town to town "like a traveling circus" undermined its reputation, and he urged the society "to cease its coquetting" with various suitors "and get married to some respectable borough with future prospects."[57]

In 1884 the society petitioned the legislature to appropriate $100,000 to purchase a fairgrounds, and the General Assembly responded by allocating half that amount, contingent upon the society raising an additional $50,000 in donations. Several cities jockeyed to become the fair's permanent home, but Des Moines, which had been home to the fair since 1879, undeniably held the inside track. After months of wheedling and pleading, the society amassed $50,000 in pledges from railroad companies and Des Moines businessmen and purchased the hillside farm of Calvin Thornton east of Des Moines as the site of the Iowa State Fairground. Throughout the spring and summer of 1885 the farmstead clattered and rasped with the sounds of hammers and saws as the agricultural society, businesses, and civic organizations scrambled to construct more than fifty new buildings on the grounds. Surveying the site in the summer of 1885, John Shaffer proclaimed excitedly that the former Thornton farm had been transformed, and that "everything denotes a *Fair Ground* [emphasis in the original]." Creating a permanent fairgrounds undoubtedly ranked at the top of the society's achievements in its first three decades. But acquiring a home for the fair did not solve all of the agricultural society's financial problems, and the long-running debate over the fair's proper role remained far from settled.[58]

2 Carnival in the Countryside

In the fall of 1854, Louisa Parker wrote to the *Iowa Farmer* to register her disappointment that the upcoming inaugural Iowa State Fair offered no prize for female equestrianism. The recent Ohio State Fair, she pointed out, had awarded a gold watch to the best female equestrian, and "the spectacle of ladies contending for the premium was the most attractive feature of the whole Fair." Parker was eager to take the reins and compete for the title of Iowa's premier female equestrian and claimed to know of several other women who also wanted to enter the contest. State Agricultural Society president Thomas Clagett replied that the new society's meager treasury did not permit it to offer premiums for such an event but gallantly offered, at his own expense, to award a gold watch "to the boldest and most graceful Female Equestrian." The competition, Clagett predicted, would provide "a fair test of superior horsemanship among the ladies of Iowa."[1]

The agricultural society's officers foresaw that female equestrianism would prove the most popular event of the fair and slated the contest as the finale of the fair's three-day schedule, touting it as the climax of the entire exhibition. As the contest began, spectators pressed against the rope surrounding the fair's track as ten women, "splendidly arrayed in long and sweeping riding habits, with feathers and ribbons to match," each accompanied by a mounted cavalier, rode onto the track. After a brief admonition from the judges about proper, ladylike horsemanship, the contestants took turns riding their horses around the course, to the delight of the huge crowd.[2]

Spectators cheered lustily for their favorite riders. To mask the contestants' identities, they competed anonymously, identified to the audience only by the color of their riding habits. A writer for the *Iowa Farmer* declared the riding of the crowd's overwhelming favorite, "Broad Blue Ribbon," thirteen-year-old Eliza Jane Hodges of Iowa City, "the most dashing, terrific, and perfectly dare-devil performance ever witnessed on horse-back. The scene was thrilling, fearful—magnificent!" Hodges, "mounted on her proud and untamable charger . . . flew around the course with the rapidity of lightning and with the sweeping force of a whirl-wind. And all this with a childlike smile upon her countenance, and her whip in full play!" When all the contestants had completed their rides, the crowd roared its admiration for the riders "and bowed to the magic of beauty." The *Iowa Farmer* summed up the event with journalistic understatement, calling it "the most thrillingly interesting and sublimely beautiful spectacle which has ever been presented within our borders, if indeed it has ever been equaled in the history of our country."[3]

The chairman of the judging committee, smitten by the young ladies' exploits in the saddle, saluted the contestants for "performances which we shall all remember as among the most pleasing reminiscences of the past," and for adding "a new wreath to the brow of beauty which already adorns our State." After a brief deliberation, the judges awarded the gold watch "for the most bold and graceful riding" to the rider adorned in the pink ribbon, Belle Turner of Lee County. The crowd, however, so vociferously preferred the breakneck riding of young Eliza Jane Hodges that, within minutes, spectators had donated nearly $200 on her behalf, as well as scholarships to girls' schools (as "she is but a child, poor, and unlettered") in Fairfield and Mount Pleasant.[4] Upon sober reflection, the *Iowa Farmer*'s correspondent conceded that the committee's decision, although unpopular with the crowd, "was based upon correct grounds." If Iowa's women were to adopt "a tasteful, correct and ladylike style or school of lady equestrianism, such as we should be willing that our wives or daughters should imitate," fairs must set an example by rewarding proper horsemanship, not daredeviltry. The reporter's initial enthusiasm, however, was truer to the spirit of the event, which was intended to entertain, not to cultivate equestrianism or etiquette.[5]

The following year, the society added female equestrianism to the fair's official premium list, justifying its inclusion by explaining that agricultural societies had a duty to foster improvement in Iowans' horsemanship. The premium list sternly warned contestants that the competition was not intended "to encourage ladies to train themselves for the Circus, or to perform daring feats of horsemanship," and that the judges would not tolerate "'break neck' or otherwise daring riding." Instead, the fair would permit only "graceful, easy riding, such as may be practiced in our cities, in our towns, on our high ways, without danger or fear, and with perfect regard to graceful and healthy exercise."[6]

The contest's spectators evidently did not share the agricultural society's high-minded goal of promoting proper horsemanship, but they did enjoy gawking at the women riders. Mary Hanford, writing in the *North-Western Farmer*, informed her female readers that women's equestrianism served no beneficial purpose and implored would-be contestants not to "place yourself in a position to have your form and carriage commented upon by a motley, gaping crowd." A correspondent for the *North-West* confessed that he found the equestrian contest exciting, but fretted that "its tendency is to draw woman from that retiring delicacy of character, and gentleness of demeanor, which are her chiefest charms in every relation of life."[7]

The most strenuous objection to female equestrianism came from the agricultural society's secretary, John Wallace, who was concerned less about its impropriety or inadvertently advancing the cause of women's equality than about "one part of the exhibition swallowing up the balance." Wallace acknowledged that female equestrianism attracted huge crowds but worried that the event's extraordinary popularity distracted fairgoers and journalists alike from the fair's true object, promoting scientific agriculture. One of Iowa's most accomplished horsemen, Wallace denied that female equestrianism served any economic purpose in the Midwest, which was "not a country in which ladies ride on horseback." Equestrianism was an entertainment, and a risqué one at that, which exposed young contestants "to the coarse ribaldry and obscene jests of a thousand 'rowdies,' who are always present on such occasions." As the fair's secretary, though, Wallace found himself torn between his misgivings about female equestrianism and the need to attract a crowd to the fair. Despite his opposition to the contest, he conceded that the fair "must have some excitement, or the institution will become bankrupt," writing that "there must be something to attract and please the crowd, in order to enable you to meet your expenses." If patrons truly attended the fair to learn, no entertainments would be necessary to lure them to the fairgrounds. Unfortunately, as Wallace estimated, "not more than one in fifty goes to improve his judgment, or increase his knowledge," so "some inducement must be offered to bring out those who merely go to see what is to be seen in the way of excitement."[8]

At Wallace's urging, the society dropped female equestrianism from the fair in 1857, but the event occasionally reappeared in the fair's official program or was conducted on the fairgrounds under private auspices over the next two decades. The "ladies' race" at the 1880 fair confirmed the darkest fears of those who charged that female equestrianism appealed primarily to spectators' prurient interests. Despite the society's attempt to maintain decorum by stipulating that contestants "be required to ride as ladies usually do; i.e., not astride," the contest seems to have been star crossed from its outset. An enormous prize of $300 had enticed only one entrant, professional rider Nellie Burke. At the last minute a Miss Boyd from Des Moines gamely volunteered to rescue the contest from default by competing against Burke. As the *Iowa State Register* tersely put it, "It would have been better if she had not."[9] Fair officials hastily improvised a riding habit for Boyd to wear. Her mount, Jim Murphy, proved so high-spirited that she struggled to control him, stay in the saddle, and keep her riding habit in place. She managed to hang on to the reins, but not to her costume, which took its own course, "exposing her person in a most disgusting manner." The fair's marshals struggled to bring Jim Murphy under control and lead him to a nearby stable, where Boyd at last found refuge from the jeering crowd.[10]

A few years later, Secretary John Shaffer flatly rejected an offer by William Perry of Moorville to stage an exhibition of female equestrianism, featuring Perry's daughters, at the upcoming fair. Shaffer advised Perry to "learn your daughters some other art than riding horses for the amusement of the public at Fairs. It will degrade them sooner or later. I have seen the evil effects of it, and know whereof I speak" (Shaffer, who had carried on a dalliance with professional rider Nellie Burke, was well acquainted with the female equestrianism's degrading effects.)[11]

The pitched controversy over female equestrianism typified the debates that surrounded state and county fairs for decades, as fairgoers, agriculturists, and journalists pondered what events deserved to be included in or excluded from the annual exhibition and what activities legitimately contributed to progress in farming, stock raising, manufacturing, home economics, and other useful arts. Could female equestrianism plausibly be justified as aiding the state's development, or was it strictly an amusement—and a titillating one at that? More broadly, did amusements invariably subvert the fair's educational mission, or could some entertainments offer harmless diversion? Could entertainments contribute indirectly to the fair's educational mission by attracting patrons to the fair? Could entertainments actually prove beneficial to hard-working farmers? Was it even possible to draw a clear line between the fair's educational exhibits and its amusements? Iowans invariably mulled over these questions as they strolled the fairgrounds, surveyed prodigious exhibits of crops and animals, cheered on race-horses barreling down the homestretch, or slipped furtively under a tent flap to gawk at a sideshow.

Iowans' debate over the role of amusements at the fair was part of a nationwide discussion of the relationship between work and leisure in the second half of the nineteenth century. As Daniel T. Rodgers points out, as industrialization transformed the American economy in the decades after the Civil War, growing abundance undercut the importance formerly attached to work and production as "the highest goals of life," and some Americans began to believe that leisure and play were not frivolous or harmful but essential.[12]

In Iowa, the annual state fair figured centrally in the debate about work and play. Many agriculturists maintained that the fair ought to be devoted to science and productivity, but others gradually, and sometimes grudgingly, conceded that a fair without amusements would be sparsely attended. At their annual meetings, in their publications, and in their advertisements for the fair, agriculturists maintained a stark distinction between the fair's serious purposes and its entertainments. On the fairgrounds, however, the two could not always be kept distinct. The fair often proved an unruly event, one that could not always be strictly governed by the society's officers and rules. Showmen, gamblers, and merchants lined the roadways near the fairgrounds and sometimes were granted admission to the grounds. Many agriculturists conceded that shows, games, and horse races attracted patrons to the fair, and that without these diversions, the fair would almost certainly sink into bankruptcy. Secretary Joshua Shaffer acknowledged that the fair ought to amuse Iowans, as well as instruct them, when he described the fair's myriad purposes in 1869:

> The fairs are the instrument for calling the people together, for relaxation from the toils of the farm, the work-shop and counting-room. They are a book out of which every visitor may read something useful and instructive. They are the special occasion of relating experiences and comparing observations. They are a market where stock, and implements, and grains may be bought or exchanged. They are a potent agency in the cultivation of the social element; and moreover they afford rational amusement, and give a holiday to the overtaxed brains and muscles of a people, the tendency of whose life is to neglect the development of their powers in the direction of ease and comfort and to allow the whole being to be absorbed in gain, of wealth, place, or fame.[13]

The task of organizing and hosting the annual state fair, John Shaffer had observed, required agriculturists "to draw numerous lines of distinction," determining which items to include in the premium list, how to judge them, and what entertainments were permissible at the fair. Agriculturists drew lines—on the fairgrounds, in the premium list, between exhibits and entertainments—but maintaining these boundaries in practice was difficult. Throughout the nineteenth century, some agriculturists insisted that shows, games, and dubious salesmen had little or no place at a scientific exhibition, and they sought to exclude showmen from the fairgrounds. In an effort to demarcate the fair's scientific mission and reputation, agriculturists cordoned their exhibition from the showmen, merchants, and other hangers-on who flocked to the fair to play to capitalize on its crowds, but fixing the boundary between legitimate agricultural exhibits and entertainments proved an elusive task, in part because these two aspects of the fair sometimes overlapped. Although agriculturists tried to divide the fair neatly into agricultural exhibits and entertainments, exhibits were arranged to catch viewers' eyes, livestock were shown in a ring, and amusements and sideshows could certainly furnish instruction in the ways of the world. The line demarcating education and entertainment was not always clear, and both were integral to the fair.

The Equine Race

The fair's horse exhibits, especially its races, forced agriculturists to confront the fuzziness of the boundary between exhibits and amusements. Agriculturists, farmers, and journalists could all work themselves into a lather over the proper classification and judging of cattle and hogs, but they reserved their real fury for the fair's horse competitions and races. Horse races indisputably ranked as the fair's most popular attraction in the nineteenth century, especially for men, and races were crucial to the fair's success. But controversy erupted over horses' importance to the state's agricultural economy, over the propriety of horse racing at the fair, and over horse racing's inevitable sidekick, gambling. Disputes over horse racing at the fair compelled Iowans to confront the blurred boundary between agriculture and entertainment.

Horses were classified second, after cattle, in the fair's earliest premium lists, in accordance with their indispensable value as draft animals and as a means of transportation. No one disputed that agricultural societies ought to offer prizes for draft horses bred to pull the plow or

the family wagon, although some agriculturists and farmers contended that horses ought to receive much smaller premiums than cattle and
hogs. But the fair offered premiums not only for draft horses but for
trotters, pacers, and racers, which provoked bitter disputes. Some critics insisted that fast horses would not speed economic development,
and that horse racing would invariably lead to gambling and corrupt
the fair. Recounting the races at the 1857 fair, the *Muscatine Weekly
Journal* charged that gambling was rife at the racetrack and declared
that "if the interest of fairs can only be kept up by exercises of this kind,
they had better be abandoned." Many agriculturists staunchly opposed
awarding premiums to racehorses, but the fair's secretary typically rationalized horse racing by pointing out that speed was a desirable at-
tribute in horses, and, more pragmatically, that without races the fair
would go broke.[14]

Horse races were the fair's biggest draw and often eclipsed its agricultural exhibits. After the 1860 fair, the *Iowa City Republican* lamented
the "meagre" exhibit of draft horses and contrasted the paltry attendance at the horse exhibit and plowing match, in which farm youth
competed to plow the fastest and straightest furrow, with the cheering
throng pressed against the outside rail of the fair's track. The newspaper noted that a mere handful of "really practical men" scrutinized the
exhibit of workhorses, while "the great mass" were not as interested "in
the efforts made to 'speed the plow' as they were in the efforts to 'speed
the horse.'" The newspaper sternly reminded its readers that "It is *the
furrow* and not the *course* which pays in this or any other State, and he
who can give to our people the best workhorses, does more for our real
interests, than as though he had secured to our State all the glory of all
the turf-stock in the country [emphasis in original]."[15]

At the fair, however, the racecourse *did* pay, and races, whether officially sanctioned or tacitly condoned, remained a vital part of the fair's
program for decades. The racetrack was the most prominent physical
feature on the fairgrounds, and the fair's schedule seldom failed to include trotting and racing, which were indispensable to the exhibition's
success. From 1854 to 1862, the premium list's horse division included
classes for roadsters and thoroughbreds but omitted pacers and trotters. At the agricultural society's annual meeting in 1859, agriculturist
Suel Foster attempted to check the growing popularity of horse races,
introducing a motion "to regulate the trotting and traveling of horses
upon the course of our Fair Grounds, so as to exclude it entirely a great
portion of the time, that it may not be a constant attraction of the peo-

ple's attention." Foster's proposal provoked angry debate, after which the delegates rejected his motion and entrusted the fair's officials to keep racing "within due bounds."[16]

Within a few years trotters commanded by far the largest premiums of the fair's many exhibits, and the number of races grew steadily. Many critics, and even a few fans, of racing charged that the growing popularity of trotting matches and races encouraged gambling, and that the races' outcome was often fixed, both of which were true. Fairgoers and journalists alike complained of rampant illegal wagering at the fair and accused professional gamblers of fixing races by paying jockeys to stiff their horses. According to the *Dubuque Democratic Herald*, the trotting match between Tom Hyer and Naboclish at the 1863 fair "was a great farce and humbug, and that is all that can be said about it." The following year, the state legislature banned horse racing, along with liquor selling and gambling, inside the enclosure or within one-half mile of county fairgrounds, although legislators conspicuously neglected to extend this ban to the state fair.[17] Dismayed by the popularity of horses and horse racing, the society's president, George G. Wright, reminded his colleagues in 1865 that fairs "are not and should not be *horse shows* [emphasis in the original]." He conceded that horses merited generous premiums but warned that "this department must not overshadow all others." Races, he declared, were a Faustian bargain, attracting patrons but causing the fair to "lose immensely in public estimation."[18]

Some agriculturists sought to justify racing by pointing to horses' economic utility as draft animals. The agricultural society insisted somewhat flimsily that it rewarded racehorses' overall excellence, not merely their speed! The "General Rule Governing Speed" in the 1874 premium list stipulated that "in all classes where the terms 'fast and far' and 'speed' are used, it is the intention of the Society not to encourage speed alone, but speed, style, size and general endurance of the animals. The committees shall, in all cases, test the qualities named." Other agriculturists excused racing as an expedient, which boosted the fair's attendance and paid for its educational exhibits. To those who charged that races diverted patrons' attention from the fair's agricultural displays, the society countered that without racing there would be no agricultural displays, indeed, no fair at all. In 1871 a committee of society members appointed to study the fair's races and attractions concluded that "it is the dictate of sound prudence and wisdom to offer at our fairs some inducement for the attraction of the multitude. Coming, perhaps, to see a trot they replenish the treasury, while they learn other excellent things from the articles on exhibition, without knowing it, and hence are profited." Horse racing might be an expedient or even a necessary evil, but it furthered the fair's agricultural mission.[19]

While many agriculturists considered races necessary to entice patrons to the fair's agricultural exhibits, their underlying motive for permitting trotting and racing was even more pragmatic: races kept the fair in the black. Secretary John Shaffer replied to a foe of racing in 1878 that "I am no horseman, that is when it comes to racing. I would not walk across the street even to see 'Racers' trot but there are thousands who would, and I have no desire to prevent their enjoyment. The American people want excitement and if we can give it to them at our Fairs in the form of Trotting and Running purses (without pool selling) let them have it."[20] When the agricultural society was chronically strapped for funds in the 1870s and early 1880s, Shaffer became downright unapologetic about racing: "It is often said that too much is devoted to trotting," Shaffer wrote, "but the TROTTING HORSE PAYS HIS WAY, and is about the only thing that does. Whether the State Fair would be a success without speeding is to be tried [emphasis in the original]." Racing's critics protested loudly, but its supporters cheered even louder, and the agricultural society never made the risky bet of hosting a fair devoid of horse races.[21]

Racing's popularity threatened to outstrip the fair's other exhibits. In 1865 horses overtook cattle at the head of the premium list, and the amount of premiums offered for trotters surpassed those for workhorses. In 1869 the fair's horse competitions included several premiums for speed; by 1874 the schedule included an extensive schedule of races, offering premiums totaling $1,700. By 1882 the horse division had been expanded even further and now comprised classes 1 through 37 ½ in the fair's premium list, with classes 18 to 37 ½ assigned to racing. "Speed" and "Horses" soon became separate departments in the premium list. Despite perennial controversy over racing, it had gained a measure of acceptance. As the *Iowa State Register* reasoned in 1881, "Horse-racing may be an evil, but until human nature is reconstructed, it will be an evil largely patronized by good and bad people indiscriminately, male and female, clergy and laity, professional and non-professional. Since, then, we must have it, it is gratifying to have it at its best."[22]

After years of debate, the propriety of agricultural societies offering premiums for "trials of speed" literally came to trial in 1881, when the Iowa Supreme Court struck down the state's prohibition of horse racing at county fairs, ruling that horse races were permissible because Iowa

law required agricultural societies to offer premiums for the improvement of livestock. According to the court, although horse races per
se were illegal in Iowa, the law did not prohibit agricultural societies
"from allowing trials of speed or horse-racing as a means of improving
the stock of horses." The court's legal hair-splitting was perhaps as clear
a verdict as could be expected, given racing's place astride the boundary between fairs' agricultural exhibits and entertainments.[23] Horse
racing provoked controversy precisely because it forced agriculturists,
farmers, and journalists to confront the complicated relationship between agriculture and entertainment at the fair, and to recognize that
the boundary between the two could not always be drawn clearly. Some
agriculturists claimed that races furthered legitimate agricultural and
economic purposes, while others excused races because they lured
patrons to the fair and to its exhibits. Both supporters and detractors
agreed on one thing: races undeniably provided the fair's most popular
and lucrative attraction.

Without the Enclosure . . .

In addition to its horse races, the fair offered plenty of other attractions, some of which neither agriculturists nor fairgoers deemed even
remotely educational. But the fair often proved unruly and could not
always be confined within the bounds of propriety. Fairs had been occasions for entertainment and for the exchange of goods and information
since the Middle Ages, and shows and games had accompanied European and American fairs for centuries. The state fair's entertainments
compelled Iowans to puzzle over the role of consumption and entertainment at a fair devoted to production and education.

In the 1850s and 1860s, agriculturists kept most amusements literally beyond the pale, that is, outside the walls of the fairgrounds. When
preparing for the exhibition, the society erected a high fence around
the fairgrounds, not only to ensure that no one entered the fair without
purchasing a ticket but also to separate, physically and symbolically,
the activities of the fair proper from the scads of showmen, merchants,
and other hangers-on who gravitated to the annual exhibition. Some
showmen and merchants were admitted within the enclosure, but
most pitched their tents along the roads leading to the fairgrounds,
where they cajoled money from the fair's patrons both coming and going. Many of the "sideshows" (so named because they were kept outside
the gates and were not officially considered part of the fair) that con-

gregated around the fairgrounds offered games of chance, exhibits of human and animal curiosities, or small vaudeville acts. Accompanying these shows were scores of food and beverage vendors, itinerant peddlers, and others eager to profit from the huge crowd attending the fair. "Without the enclosure" at the 1856 fair, a reporter observed that "various showmen, with voice and music, were trying to attract the public attention to the 'Educated Alligator,' 'the California Bear, with but two legs,' and 'The Living Skeleton.' Even the hand-organ and the monkey were having their full share of attention." At the 1860 fair, one newspaper reported that "outside the enclosure, all through the day and late into the evening, there was fun which flew 'fast and furious.' The usual accompaniments of Fairs were all in place and doing an excellent business." After the fairgrounds had fallen dark and silent, the crowds outside continued their revelry long into the night, "seeing how near to the skies the 'celestial rail road' would carry them, or in some other way striving to 'drive dull care away.'"[24]

The fair's enclosure could not always prevent entertainers from injecting some unsanctioned levity into the exhibition. Fair officials were annoyed in 1855 when a group of young men in "Calathumpian" disguise, billing themselves as the "Chinch Bug Guards" and "Earthquake Volunteers and Flying Artillery," rode their mounts onto the track and proceeded to delight and irritate the audience with their antics. While some fairgoers enjoyed the impromptu show, others denounced it as "idiotic tomfoolery" that undercut the dignity of the exhibition. "Captain Sky High," the Calathumpians' ringleader, retorted that "the mission of the company was accomplished, and those who were 'really wise, refined and religious'" had enjoyed the show. The Calathumpians' uninvited and carnivalesque performance briefly upended the exhibition's decorum, turning the fair topsy turvy.[25]

The role of entertainment at the fair provoked controversy for decades. Chartered to promote scientific farming, the agricultural society permitted amusements at their exhibitions only after considerable debate, and over the vehement objections of some of its members. Financial considerations impelled agriculturists to admit amusements to the fair, but their educational pretensions sometimes made it difficult to justify frivolous shows and games. Many agriculturists rationalized the inclusion of amusements because they lured people to the fair, where they would almost certainly spend some time perusing the agricultural displays. By the end of the nineteenth century, however, a growing number of the society's officers viewed amusements not as an expedient or a necessary evil but as a tonic for the state's hard-working farm families. As Iowans debated the respective place of agriculture and entertainments at the fair between the 1850s and the turn of the twentieth century, they were often implicitly discussing the future of their state. The uneasiness or outright antipathy that some agriculturists, journalists, and fairgoers felt toward entertainments was not merely a cranky outburst against the prospect that someone might actually have fun at the fair. To many agriculturists, amusements represented the opposite of everything that the fair officially existed to promote. Instead of productive labor and scientific agriculture, shows and games represented frivolity, consumption, even outright fraud.

Although the leaders of the state agricultural society considered it an organization that served the public good by fostering education and economic growth, its annual fair was a boisterous and ephemeral event. The itinerant showmen and merchants who pitched their tents on and around the fairgrounds were the most transitory participants in an already disconcertingly fleeting event, and their shows, games, and products subverted the society's efforts to conduct a meticulously organized exhibition of scientific agriculture. Showmen sometimes made a mockery of the fair's scientific exhibits with their displays of human and animal freaks and other curiosities, while the hordes of salesmen and gamblers plying their trade at the fair offered society members and fairgoers an object lesson in the unruliness and even unscrupulousness of the market economy. Showmen and merchants kept up a dissonant clatter around the society's attempt to cordon off an orderly arena devoted to scientific and economic progress.

Just as the society strove to adhere to an orderly schedule in its agricultural competitions, it also attempted to impose order on showmen and merchants, who proved far less susceptible to regulation. Troubled by the sideshows and stands that annually clustered around the fairgrounds, the society complained in its report for 1857 that "on occasions of State Fairs, great numbers of vagabonds always assemble to make their money by their wits; by stealing, gambling, or some other almost equally objectionable method." So long as showmen and merchants of any sort were allowed in the vicinity of the grounds, the society complained, liquor, gambling, and freak shows would invariably surround the fair, "greatly to the annoyance and injury of the Exhibition, both in its morals and its receipts." The society resolved that, in the future, it would either compel showmen and merchants to purchase a license and work within the fairgrounds, which would increase the fair's revenues and afford the society greater control over sideshows, or force them to leave altogether.[26]

Many of the salesmen hawking their wares at the fair made a sham of the fair's scientific pretensions by selling patent medicines or shoddy merchandise, and every year the society's secretary rejected stacks of salesmen's requests for permission to sell their wares on the grounds. Shortly before the 1871 fair, Secretary Joshua Shaffer wrote in his diary that his mailbag was filled with dozens of "the usual queries—from editors, sewing machine men—hucksters—thimble-riggers and confidence people." During the 1872 fair, the streets of Cedar Rapids teemed with "the ARABS OF COMMERCE plying their deceptions upon the street corners [emphasis in the original]."[27] Patent medicine salesmen touted remedies guaranteed to cure ailments ranging from bunions to migraines. A few of these remedies may have been effective, but most were quackery, if not downright harmful. A reporter at the 1869 fair observed "venders of tooth powder with poor teeth; owners of respirometers with small chests; magnetic battery operators who look sick and dyspeptic." These salesmen frequently requested permission to operate on the fairgrounds, but agriculturists generally deemed them unworthy to participate in an exhibition devoted to the promotion of science. When patent medicine salesmen did gain admission to the enclosure, the society relegated them to space alongside the shows, at some remove from the fair's agricultural exhibits.[28]

In addition to hordes of itinerant shows and salesmen, the fair occasionally contended with more formidable rivals, who sought to poach from the fair's crowd. Traveling circuses sometimes played havoc with fairs across the Midwest by deliberately booking engagements to coincide with state and county fairs. In 1873 the state fair was beset by J. A. Bailey's Grand International Menagerie, Museum, Aquarium and Circus. Upon learning that Bailey's company planned to pitch its tents in Cedar Rapids during the fair, and in neighboring towns before and after the exhibition, Secretary Joshua Shaffer exclaimed that "we have fallen into the hands of the Philistines. 'Bailey's Big Show & Circus'—a monster as big as Barnum's." Agriculturists drew a sharp distinction between their livestock exhibitions and the circus's exotic animals and lavish spectacles, and they resented circuses for siphoning off the fair's patrons and profits. But Shaffer also recognized that taking a few tips from showmen could add to the fair's appeal and coax crowds onto the fairgrounds.[29]

Here They Come

Showmen, salesmen, and charlatans could be a nuisance, but other hangers-on posed more difficulties for the fair. The annual state fair was the only occasion that attracted a crowd of thousands in the pioneer era, and pickpockets and swindlers often traveled to the exhibition to prey upon fairgoers. Held in one of Iowa's cities and thronged with visitors and exhibitors from around the state and the region, the fair offered its visitors an experience utterly unlike the quiet and isolation of small town and rural life. While the fair generated pride and profits for local residents, the crowds and commotion surrounding it sometimes overwhelmed the host town. Maintaining order on and around the fairgrounds was indispensable to the agricultural society's effort to promote an image of economic stability. The fair's publicity often boasted that the exhibition was as sober and decorous as a church picnic, but the enormous crowds, flamboyant showmen, and even outright criminals who attended the fair posed a serious challenge to the society's effort to cordon off an exhibition that presented a microcosm of a well-ordered agricultural economy.

Newspapers frequently admonished residents and fairgoers to beware of the crooks who attended the fair to rob or cheat unsuspecting patrons. "LOOK OUT," the *Dubuque Democratic Herald* warned local citizens in 1863. "Beware of thieves, pickpockets and swindlers. They are arriving on every train (especially from the east) and throng the city. . . . This class of men follow fairs and public assemblages to ply their vocation and practice their arts upon the unwary [emphasis in the original]." The 1869 fair in Keokuk was beset by more than its share of troubles. A local newspaper reported that "a number of little unpleasantnesses occurred," including several concerted efforts to crash the fair's gates, along with the usual thieving and swindling. One miscreant was jailed for firing his pistol at the local police chief, while "two or three men were stabbed, and one woman had her arm broken" near the fairgrounds. Four years later, the *Cedar Rapids Times* complained that burglars "held high carnival on the grounds and throughout the city" during the fair, adding that local residents were "not sorry that Fair week is over."[30]

The 1875 fair in Keokuk, according to the *Daily Gate City*, attracted "the usual quota of that nomadic and degenerate class usually attendant upon occasions of this kind." The newspaper reported that criminals had infested every town for miles around Keokuk in advance of the fair. A correspondent from Farmington wrote that "even our quiet little city, fully thirty miles distant, was last evening the scene of several robberies perpetrated by scamps on their way to the Fair." As the fair's opening day approached, the newspaper's headline warned ominously:

HERE THEY COME.
The Thieves Have Commenced
Skirmish for Plunder

According to the newspaper's sensational account, many of the criminals descending on Keokuk were members of a gang based aboard a gunboat moored in the Mississippi River. Another mob of fifty "rowdies" was plotting to take advantage of unsuspecting farmers, and more crooks arrived on virtually every train. The paper advised citizens to prepare for fair week by taking the following precautions: "Deposit your money and valuables in safe places, fasten your windows, diet your dogs to a condition of ravenousness, and above all keep your revolvers in good working order." The newspaper's fears were not altogether unfounded: during the fair's run, 126 people were arrested for disorderly conduct, assault, theft, and other offenses. "Citizens have been taught to dread fair week," wrote the *Iowa State Register* in 1884, "on account of the long list of crimes usually enacted by the gang of roughs in attendance at such gatherings."[31]

Showmen of the Higher Sort

In the fair's earliest years, its officially sanctioned entertainments were limited to horse racing, female equestrianism, and a brass band, while sideshows and games were kept outside the fairgrounds. In the 1860s, the fair's secretary began to grant a few sideshows the "privilege" to operate on the fairgrounds in exchange for a fee. Some agriculturists opposed blurring the boundary between the fair's agricultural exhibition and entertainments by allowing sideshows on the fairgrounds. George G. Wright, the society's most fervent critic of racing and entertainments at the fair, staunchly objected to allowing sideshows inside the fair's gates, worrying that shows and games would attract people who had little or no interest in learning about agriculture. At the society's annual meeting in 1865, Wright delivered a jeremiad to his peers, warning them of a movement afoot "to divert our Fairs from their true and original object and purpose . . . to be fit and just representatives of the wealth, progress, genius and enterprise of the State in all its industrial departments." Wright declared that the fair embodied Iowa, which "will be judged more or less by the character of this, one of its institutions, protected and fostered as it is, by State aid, and the money of the people." He conceded that the fair needed to turn a profit but insisted that it could succeed without pandering to the public's debased taste for entertainment. Amusements would only backfire, he warned, by attracting the "wrong" sort of patrons, corrupting the fair and alienating those Iowans who attended the fair to become better farmers and homemakers.[32]

Wright insisted that agriculturists could sustain the fair's popularity over the long term only by hosting "a *Fair in its legitimate and just sense*—and not by bringing into our enclosure shows, games and entertainments, demoralizing in their tendency and only calculated to divert the attention of the crowd from the higher aims and purposes of our exhibition [emphasis in the original]." He exhorted his colleagues to stake out the "high ground" by excluding all shows and games from the fair: "We should have but one exhibition upon our grounds. We cannot sustain an institution like ours, if we have to rely upon the thoughtless, inconsiderate throng, who think more of monkeys and human and animal curiosities than of the great moral, economical and practical lessons taught within our halls and scattered in the richest profusion over our grounds."[33]

In the 1870s, the agricultural society's officers began not merely to tolerate but to book shows, games, and other amusements for the annual exhibition. Tensions swirled around the fair's agricultural exhibits and its entertainments for decades, as agriculturists, farmers, and fairgoers pondered the fair's role and the progress and prospects of their society and culture. Despite considerable opposition to shows and games, by 1870 the fair earned several hundred dollars annually from the sale of "privileges" and "concessions" to sideshow operators. Sideshows were not allowed inside the fairgrounds without reservation, and agriculturists continued to distinguish between the fair's educational mission and its amusements by assigning showmen space far away from the fair's agricultural exhibits. It was not far away enough for some agriculturists, and after an acrimonious debate at its annual meeting in 1870, the society voted to exclude "side-shows of every description" from the upcoming fair. Only three days before the fair opened, the society's officers panicked that it was doomed to lose money and abruptly rescinded the ban. Secretary Joshua Shaffer frantically wired a batch of last-minute telegrams to sideshow operators as he scrambled to ensure that the fair would not be devoid of shows and amusements. Shaffer understood that the fair served an educational purpose but was also in the show business. He grumbled that his colleagues' sanctimonious determination to bar amusements from the fairgrounds threatened the exhibition's profits and even its survival.[34]

Many Iowans conceded that the fair ought to offer some entertainment but insisted that those amusements should not debase Iowans' morals or taste. The *Iowa Homestead and Western Farm Journal* conceded in 1873 that the fair must offer entertainment, but the magazine insisted that the fair was not simply a carnival and ought to book only wholesome amusements. "The management of a State Fair is something like the management of a huge show," it declared, "but the managers require to be showmen of the better and higher sort, for while ordinary shows are proverbially accompanied by all manner of iniquities, and are accepted as inseparable from these, our agricultural fairs are the exponents of all the useful arts and industries of the age in which we live."[35] Likening the fair's managers to showmen of *any* sort, though, acknowledged a significant change in the agricultural society's role. Agriculturists had made pronouncements about the fair's lofty purpose on the fairgrounds and at the agricultural society's meetings since the first fair. But a growing number of agriculturists now conceded that the fair could simultaneously fulfill its educational mission and offer diversion to its patrons.

Debates over amusements at the fair invariably grew louder when the nationwide depression ravaged the agricultural economy beginning in 1873. Economic downturns simultaneously inspired a resurgence of agrarian rhetoric extolling the fair's educational mission and tempted the fair's secretary to turn a blind eye to dubious entertainments to attract a crowd. When the depression drove the fair deep into debt, some agriculturists argued that shows were too expensive and that the fair should shun frivolity during hard times.[36] Desperate to turn a profit, however, the fair's secretary felt pressured to book more amusements, including games of chance, to lure patrons to the fair. Still, after the 1873 fair lost money, Joshua Shaffer, who had organized the first two fairs and overseen the fair since 1863, relinquished his position as the society's secretary and was replaced by his nephew, John R. Shaffer.

The society's new secretary, like his predecessor, was willing to allow more amusements and games at the fair as the depression dragged the fair's finances deeper into the red. In the summer of 1875, several sideshow men requested permission to bring roulette wheels and other games to the upcoming fair. Although the society's regulations prohibited gambling at the state fair, John Shaffer found himself torn between his concern for the fair's reputation and its finances. He fretted about adverse publicity but confided to the fair's superintendent of privileges that "we want the money."[37] Ultimately, the lure of money trumped Shaffer's ethical qualms: "Games of chance are numerous on the grounds," reported the *Daily Gate City*, including wheels of fortune, chuckaluck (a dice game), and a "drawing" stand (another dice game, which another local newspaper described as "a bunko hell on a small scale"). The *Western Farm Journal* charged that the fair had deteriorated, claiming that "the admission of the sports of the race track, and the renting of space for the selling of intoxicating drinks, practicing games of chance, and the exhibition of monstrosities, has grown in proportions each year, and bears fruit in keeping with its bigness."[38]

Unfortunately for Shaffer, his gamble did not pay off, and the fair's receipts spiraled downward, despite the revenue earned from selling privileges to showmen and gamblers. At the society's annual meeting a few months after the fair, several members castigated Shaffer for his management of the fair, and the society once again resolved that henceforth all games of chance "be excluded from the grounds during the Fair, as that kind of swindling has a tendency to decrease the attendance." Shaffer publicly endorsed the society's resolution, hailing it as "a move in the right direction and one the Society will never regret," but his poker-faced support for the ban on gambling was a bluff. His primary concern was to salvage the society's finances, and he continued to allow sideshows and games to operate at the fair despite the clear objection of his colleagues. Shaffer permitted a gambler to spin a wheel of fortune at the 1876 fair until an indignant agriculturist complained, and he allowed another wheel of fortune inside the fairgrounds the following year. In 1878 Shaffer attempted to persuade the society's board that games of chance ought to be allowed on the fairgrounds in order to make the fair more profitable, contending that "I can see no more harm in the wheel than in our races." But the board steadfastly refused to rescind its prohibition against gambling, which, Shaffer complained, resulted in "a meagre demand for stand and side-show privileges" at the fair.[39]

Many Americans detested gambling, not only because they considered games of chance immoral or crooked but because gambling provoked some of their deepest misgivings about the market economy. The shadowy figure of the gambler was the incarnation of capitalist speculation, a trickster who created nothing of lasting value and tempted those who were naive or weak willed to forsake honest labor for the lure of easy money. In an era in which farmers were extremely vulnerable to the boom-and-bust cycle of the market economy, as the depression of the 1870s had made painfully clear, agriculturists' opposition toward gambling was not merely an outburst of self-righteousness.[40]

Agriculturists' fears about the harmful consequences of gambling were evident when the society passed a resolution in 1878 to curb the proliferation of crooked sideshows at the state's county and district fairs. The resolution declared that the state fair had completely prohibited gambling and yet remained profitable—neither of which was true—and decreed that county fairs that permitted gambling would forfeit their state appropriation. The resolution's tone conveys agriculturists' conviction that gambling was no petty vice but posed a dire threat to rural life: "Gambling in any form is *per se*, immoral, wicked and dangerous," the society declared. "The youth and inexperienced who attend our fairs for recreation and improvement, are not only robbed of their money, but are there taught the first rudiments of that great and seductive evil, and many are led direct from thence to adopt the gambler's profession, together with all its wicked and degrading accompaniments, and are thereby ruined and lost; outcasts and vagabonds."[41] Shaffer, generally willing to risk controversy by tolerating gambling in order to increase the fair's receipts, did not share his colleagues' moral indignation about games of chance. Publicly, he championed the society's prohibition of gambling; privately, he complained about agriculturists' sanctimony to a fellow state fair secretary. "This Society has become *too moral* to allow any games of chance [emphasis in the original]," he wrote, charging that his colleagues' pious concern about protecting "the farmers' boys" from ruin threatened to drive the fair into ruin instead.[42]

Privileges, Concessions, and Attractions

While the propriety of amusements at agricultural fairs seldom failed to provoke disagreement at the agricultural society's annual meetings, entertainments steadily became more prevalent at the fair in the 1880s and even gained a measure of acceptance from many of the society's officers. Some agriculturists remained steadfast in their opposition to amusements. Others excused shows and games as a necessary evil or a frivolous but harmless diversion. But a growing number of agriculturists softened in their opposition toward amusements and even began to advocate them as beneficial, even necessary, for the state's hard-working farm families. However, although some of the society's officers were becoming more tolerant of amusements, they did not suddenly throw open the fairgrounds gate to showmen. As they permitted more sideshows and games inside the fairgrounds, they simultaneously sought to exercise greater control over them. Defenders of amusements at the fair contended that choosing and booking shows for the fair, rather than simply allowing itinerant showmen to work on the fairgrounds, could actually make the exhibition cleaner and more manageable.

Hard times in the 1870s had pressed the fair's secretary to admit amusements to the fair, because showmen added to the society's receipts both by paying a fee for the "privilege" to operate and by attracting patrons to the fair. But with the exceptions of horse races, female equestrianism, and an occasional brass band, the society did not pay to book entertainments until 1880, when it appointed two officers to scout and book attractions for the upcoming fair. After sifting through stacks of letters and advertisements from showmen eager to rent space on the grounds, the society paid $400 to book the fair's first major commercial entertainment, a Roman chariot race to be staged by an entertainment company from Chicago. The race proved so popular that the society booked it again the following year and for several years thereafter (the race's popularity was greatly aided by Iowan Lew Wallace's 1880 novel *Ben-Hur*). In addition to the chariot duel, the fair's program included "the musical prodigy, Little Ella," a brass band, and ladies' horse races.

Elated that the society's members were apparently shedding their opposition to attractions at the fair, Shaffer wrote to one of the society's officers a few weeks before the fair that the old-fashioned agricultural exhibition of gargantuan vegetables had become obsolete and been replaced by a more modern and diverting fair.[43] "The day of the American people is past to look upon pumpkins and smell onions," he declared. "We are living in a fast age, and attractions and humbugs are the order of the day. The bigger the humbug the more we take to it. We hope to give such an exhibition that will *please and tickle* the people [emphasis in the original]."[44]

Even though agriculturists began to book more amusements at the fair in the 1880s and 1890s, many of them maintained that entertainments were somehow grafted on to the agricultural fair, rather than integral to it. They continued to insist that the fair's true purpose lay in celebrating labor and productivity, not in affording Iowans an annual holiday. Despite indications that amusements were gaining greater acceptance at the fair, the society's rhetoric preserved the boundary between the fair's agricultural exhibits and its entertainments, and it often rationalized amusements as a necessary ploy to attract city dwellers, who were already familiar with commercial entertainment and took little interest in displays of animals, crops, and farm implements.[45]

Entertainments, of course, also appealed to farm families, who had far fewer opportunities for recreation than did residents of the state's cities and towns. In the 1880s and 1890s, some agriculturists began to hail entertainments as a necessary, well-earned diversion for the state's farm families, who needed to take an occasional break from trudging behind the plow or stoking a wood-burning stove and enjoy a bit of leisure and even festivity. In his opening address at the 1881 fair, the society's president declared that entertainment was not necessarily evil but could even make farmers and other workers more content and more productive. "In this utilitarian age," he declared, "all forms of entertainment had come unfairly into disrepute," with disastrous consequences for Americans, especially for farmers. "Can it be denied that the majority of the patients in our insane hospitals come from the ranks of our over-worked and over-strained farming communities?" he asked. Hard-working farmers needed an occasional respite, and only a killjoy would "condemn the merely entertaining portion of our annual exhibition."[46] Catering to the innate human desire for amusement and leisure ranked among "the great needs of the day in which we live," and the fair had a duty to supply this amusement, just as it had a duty to boost the state's economy.[47]

The agricultural society's most fastidious members remained unyielding, objecting to booking *any* entertainments for the exhibition and insisting that shows and games would ultimately undercut the fair's reputation and its receipts. They opposed paying a fee to book big acts but were especially suspicious of small, independent showmen and game operators, who were notoriously difficult to control. Once granted a "concession" and ensconced within the grounds, seemingly legitimate showmen and barkers might offer risqué acts or crooked games. Newspapers frequently charged that shows and games were often little more than fronts for lewd dancing, gambling, and liquor sales. Because sideshow operators paid a fee to work the grounds, the society had considerable incentive to overlook their transgressions, as long as they remained reasonably discreet.

The society's secretary walked a tightrope, struggling to maintain the fair's allure and profits while avoiding scandal. According to the *Iowa State Register*, Secretary John Shaffer slipped in 1882, and "THE NUMBER OF 'SNIDE' ENTERTAINMENTS of various characters has increased on the grounds over former years [emphasis in the original]." Freak shows, gamblers, and peddlers of shoddy merchandise all endeavored "to entice the vicious and ignorant, ensnare the innocent, and horrorize the finer feelings of decent people."[48] Stung by a barrage of criticism of the fair's sideshows, the agricultural society yet again resolved in 1883 "that all disgusting sideshows be prohibited from the Fair Grounds." But enacting a ban on disreputable sideshows at the society's annual convention proved far easier than actually excluding them from the fair, and a reporter for the *Register* found the grounds "thickly dotted with side shows of every description." But, as the newspaper conceded, "it is always so at every big fair, so people should not complain."[49]

Enterprising showmen eagerly capitalized on the fair's growing willingness to book amusements in the 1880s, inundating the secretary with advertisements for balloon ascensions, Roman chariot races, acrobats, freak shows, games, bands, drill teams, fireworks displays, bicycle races, and countless other attractions. Sifting through these stacks of advertisements and booking attractions for the fair began to occupy more of the secretary's time as the fair became an important venue for entertainers. The definition of an "attraction" encompassed virtually anything that might plausibly attract patrons to the fair: brass bands, acrobats, or even "some of the would-be presidents of the nation." The fair's main attraction in 1883 was Chief Sitting Bull and a group of Sioux Indians, who received $800 to appear at the exhibition, where they were billed as "the murderers of the gallant Custer." Exhibits of Indians and other supposedly "primitive" peoples were common at state fairs and world's fairs in the late nineteenth and early twentieth centuries. These "ethnological" exhibits, ostensibly educational, were often depicted by their organizers as a benchmark against which white fairgoers could measure their own economic and cultural advancement. "The same Indians who left their wigwams on the banks of the Des Moines river and rode their ponies out from a wilderness," boasted President William Smith in his opening address to the 1883 fair, "can now return in a palace car to a city of forty thousand"—a testament to civilization's progress over the past fifty years, in his view.[50] As Smith's remark suggests, ethnological exhibits invited fairgoers to measure civilization's progress by gawking at exotic, "primitive" peoples and drawing a contrast between the Indians' culture and their own.

The High Ground

Rumbling eastward aboard a streetcar from Des Moines toward the new state fairgrounds on opening day in September 1886, a newspaper reporter glimpsed a veritable city of tents, booths, and buildings, festooned with flags and banners, on the hillside ahead. A fairgrounds had sprung up where only a few months earlier a farm had stood. As he passed through the fair's gates, he was overcome by the heady sensation that he needed at least "a hundred eyes for seeing and a score of tongues for questions" as he marveled at the fair's astonishing array of exhibits and attractions. Magnificent cattle, horses, and hogs, curried until they "glisten like burnished metal." So many new farm implements that the Patent Office surely could not keep pace with inventors' ingenuity. Gaudy tents that tempted passers-by to set aside their scruples and plunk down a quarter to gape at the freak shows and burlesques within. Ear-splitting carnival barkers. "Rural Romeos and Juliets" strolling the grounds, enjoying a respite from the labors of farm life. "Human vultures" swindling gullible fairgoers with their crooked games and shoddy merchandise. After strolling through the bewildering "maze of exhibits" in the fair's colossal Exhibition Building, he emerged "with a vision of furniture, mantles, rare carvings, sewing machines, harnesses, stoves, musical instruments, paintings, portraits, photographs, spreads, quilts, embroidery, fans, handsome dresses and fair faces all mixed up in one variegated confusion." Daunted by the task of summarizing the exhibition's kaleidoscopic variety in print, he offered his readers instead a series, "State Fair Scenes."[51]

The opening of the permanent fairgrounds in 1886 marked a watershed in the fair's history, and some agriculturists hoped that the new grounds would enable the fair to jettison sideshows and games and become a full-fledged educational institution. The fair's publicity proclaimed that "Everything about the fair will be orderly, moral and instructive," and that games of chance and disreputable sideshows would be excluded from the new fairgrounds. At the dedication ceremony for the fairgrounds, a host of politicians, journalists, and agriculturists regaled the audience with addresses about the noble past and auspicious future of the state, the agricultural society, and the fair. The society's president recounted the fair's impressive growth over the past three decades. Peter Melendy of Cedar Falls, perhaps the state's most distinguished agriculturist, saluted the achievements of the agricultural society, culminating with the creation of a permanent fairgrounds. The society's president recounted the fair's impressive growth over the past three decades. Agriculturist Melendy saluted the achievements of the agricultural society, culminating with the creation of the permanent fairgrounds. John Shaffer, who had worked tirelessly over the past decade to organize the fair, confidently declared that the exhibition, for decades a fleeting event held on hastily rigged grounds, had at last attained permanence and respectability.[52]

The day's most grandiloquent speech was delivered by Josiah B. Grinnell, agriculturist, breeder of Merino sheep, politician, and benefactor of a small liberal arts college. Grinnell implored the agricultural society and the audience to shun amusements and consecrate the new fairgrounds to its legitimate purpose, promoting scientific agriculture. Now that the fair had a permanent home, he argued, it no longer needed to open its gates to dubious amusements but could at last be true to its original mission and rid itself of showmen and gamblers. "I would bar the gates forever to gamblers, jockeys, whiskey venders, and oleomargarine frauds," he declared, "and leave reptilian monsters, with acrobats, pigmies and fat women to the showman, Barnum. Then write over your portals, dedicated to art, animal industry and agriculture." Grinnell's plea to slam the fair's gate in the face of showmen offered a reminder that the fair's agricultural and educational exhibits still coexisted uneasily alongside its entertainments, and that some Iowans continued to wish that the fair could be made into strictly an agricultural exhibition.[53]

Despite Grinnell's call to uphold the sanctity of the fairgrounds, dozens of showmen had already streamed through the fair's portals and pitched their tents on the fairgrounds, and their spieling created a din jarringly at odds with his paean to agriculture. One newspaper criticized the society for permitting showmen, gamblers, and crooked merchants to "ply their nefarious schemes" at the fair, complaining that visitors could not view the fair's livestock and machinery departments without walking past sideshow tents festooned with banners that beckoned patrons to see the "gaudy impossibilities" within. The society's president reported that the fair's sideshows were cleaner than those of previous years but urged the society to prohibit all questionable shows and games, so that "there will not be anything on the grounds that will call the blush to the cheeks of our lady visitors."[54]

Some of the agricultural society's officers sincerely hoped that the acquisition of a permanent fairgrounds would enable the society to re-make the fair into a more sober exhibition and to reduce or eliminate amusements altogether. Even John Shaffer, who had often permitted games and shows to operate on the fairgrounds, predicted that state fairs would become more serious and educational in the near future. He informed the secretary of the Wisconsin state fair in 1888 that state fairs were becoming substantial, state-supported educational institutions, similar to schools and museums, and that "the day of balloons, wild beasts, chariot races, etc., are only a question of time."[55]

Opposition to shows and games was often fueled by classism and by the perception that entertainments appealed primarily to poorer and less educated people. In 1889 John Shaffer explained that constructing attractive and substantial buildings on the fairgrounds would enable the fair to become a "moral and high-minded exhibition," attract "a better class of people," and eliminate "the infernal *egg throwing*, ring boards, etc. [emphasis in the original]." He boasted that the agricultural society had become more discriminating in booking amusements for the 1889 fair, refusing space to "dance-houses, knife-boards, cane racks, and all combinations of show men and women." The society's president urged his colleagues to "continue to draw the boundaries tighter until all side shows are excluded."[56] Shows, he complained, were "hard to confine within the bounds of decency" and "a constant menace to the character of a fair. Shunned to-day by the larger and better classes of our citizens, they are surely not in keeping with the higher civilization it should be ours to seek."[57]

But the agricultural society's deeds often failed to match its high-minded rhetoric. The *Register* reported that many of the shows at the 1889 fair "are right up to the standard of first-class frauds." The society no longer excluded sideshows but allowed them on the fairgrounds in exchange for a fee. As Shaffer noted, "It was formerly the rule to exclude all theaters, mountebanks, jugglers, concerts, etc., from the grounds, but in such years, they pitched their tents outside, obstructing the highways, keeping up a continuous din and uproar, and were a more insufferable nuisance than when within the enclosure and under the control of the society."[58] As always, the society's secretary had a powerful incentive to book amusements, and when the fair's gates opened each September, a variety of attractions awaited the thousands of visitors who plunked down their fifty cents and streamed in.

Critics of entertainments, however, continued to insist that the fair could succeed without shows and games. In 1891 Shaffer wrote to his counterpart at the Minnesota state fair that "the time is not far distant when we shall have to abandon in a measure, the attractive feature except the Fair proper. We should educate our people that the Fair should be the attractive feature, without any 'Wild West,' balloon ascensions and etc." At the 1892 fair, George Wright, the society's former president and its most ardent foe of amusements, delivered an address advocating a "proper" fair and inveighing against sideshows. Maintaining an attractive and clean fairgrounds, Wright said, was "a matter of first importance," because unkempt, dusty grounds ran counter to the society's vision of well-tended farmsteads and economic prosperity. Tellingly, Wright seamlessly elided the importance of keeping the fairgrounds tidy with a pointed denunciation of shows and games. In his words, the fair must also occupy the moral "high ground" by excluding sideshows in favor of a genuine agricultural fair.[59]

The Year of All Years

The 1890s ranks among the most tumultuous decades in American history, during which a wrenching financial depression, strife between workers and corporations, a vast influx of immigrants, and the Populist insurgency of disgruntled farmers combined to strain the nation's political and economic system nearly to the breaking point. At the decade's outset, the 1890 Census signaled the end of the frontier era by revealing that unoccupied land was no longer available for settlement in the West. The closing of the frontier prompted widespread hand-wringing that the wellspring of American distinctiveness and democracy had run dry. At the end of the decade and the century, many Americans' faith in unbounded progress had been severely tested, if not altogether broken.

The political and economic upheaval of the 1890s could hardly fail to affect the state fair. As the *Homestead* observed, the fair "floundered about in multiform embarrassments" throughout the decade.[60] In 1893 the depression and competition from the 1893 World's Columbian Exposition in Chicago combined to plunge the State Agricultural Society deep into debt, from which it never rebounded. Hard times made it impossible for the fair to meet its expenses or even to pay premiums to its prize winners. Most important, the fair's difficulties in the 1890s only intensified the long-running debate over the fair's purpose. Disagreements over whether the fair should emphasize its educational role or offer its patrons more amusements divided both the agriculturists and Iowans generally.

The fair suffered not only from the economic depression of the 1890s but also from a gnawing sense that old-fashioned agricultural fairs, like sod houses and scythes, were fast becoming a relic of the pioneer era. From its inception, the fair had been devoted to progress and had displayed Iowans' finest achievements along with the most up-to-date agricultural knowledge and technology. At the end of the nineteenth century, however, the fair, which had once looked resolutely toward the future by heralding scientific and technological innovation, became somewhat backward-looking, even nostalgic. Fairs had always promoted progress and prosperity, but the wrenching events of the 1890s tested Americans' confidence that progress was inevitable and their faith that prosperity would one day return.

In 1893 the stupendous World's Columbian Exposition, hosted to commemorate the four hundredth anniversary of Christopher Columbus's first voyage across the Atlantic Ocean, opened in Chicago. The World's Fair dazzled some twenty-seven million viewers (in a nation with a population of roughly seventy million!) with its gleaming White City, brimming with an astonishing array of exhibits of the most advanced products of American industry, and its celebrated Midway, lined with diversions. Because Chicago lay only 340 miles east of Des Moines and less than 200 miles from Iowa's eastern counties, the World's Fair threatened to siphon attendance from the state fair and county fairs. These smaller fairs, with their dusty grounds, rustic exhibition halls, and small-time sideshows, could not hope to compete with the fantastic splendors of the Exposition's neoclassical White City or its instantly legendary Midway, teeming with entertainments and ethnological exhibits of peoples and cultures from around the globe.

State and county fair secretaries throughout the Midwest were jittery about the Chicago fair's impact on their attendance. Clearly, agricultural fairs could not remotely rival the exposition's exhibits or attractions, and many agriculturists questioned the wisdom of holding state and county fairs in 1893. The International Association of Fairs and Expositions, an organization of fair managers, resolved that no fairs should be held in the Midwest in 1893, because smaller fairs could not compete with the Columbian Exposition, and because many showmen would be engaged at the World's Fair. Public sentiment in Iowa, however, overwhelmingly favored holding the annual state fair, and the state's newspapers urged the agricultural society not to sacrifice the state fair to the Columbian Exposition. After the agricultural society voted to host a fair in 1893, John Shaffer dismissed naysayers' gloomy predictions that the fair was doomed to lose money: "It is the year of all years in which to hold a Fair," he declared confidently.[61]

Even without competition from the World's Fair in Chicago, 1893 proved an extraordinarily trying year for the state fair. A nationwide financial depression convulsed the economy in the spring, clouding the fair's prospects. Farmers felt keenly the depression's ravages, fueling the growth of the Populist movement, which had gathered strength in the 1880s. Populism, which began as a series of local and state Farmers' Alliances, tapped into farmers' economic and political grievances by advocating regulation of corporations, a monetary policy that would make credit more plentiful, and measures to increase farmers' income by enabling them to market their crops for a higher price. While the Farmers' Alliance and People's Party, founded in Omaha in 1892, were not as potent in Iowa as in portions of the South and the plains, Populism held considerable appeal for many farmers, and the fair's managers permitted the Farmers' Alliance and People's Party to host their annual meetings on the grounds during fair week throughout the 1880s and 1890s. Populism, like the Grange before it, ran counter to the boosterism espoused by the state agricultural society, and most of the society's officers viewed anxiously the growing clout of the populist movement. Shaffer warned ominously that the populism would cause "ruptures" if the movement became involved in politics, a fear that was realized with the creation of the People's Party.[62]

As the 1893 state fair approached, the fair's publicity, as always, remained extravagantly optimistic; privately, however, the society's officers fretted about impending financial disaster. Competition from the World's Fair, along with the deepening depression, led many of the society's officers to cast about for some gimmick to boost the fair's attendance. The fairgrounds had been wired for electricity the previous year, and John Shaffer predicted that the marvel of electric lighting and the "nocturnal splendor" of nighttime entertainments would enable crowds to remain on the fairgrounds well into the evening. But the fair faced stiff competition from Des Moines businessmen, who since 1889 had sponsored a street festival, Seni-Om-Sed (read it backward), sched-uled to coincide with the fair. In 1893, Seni-Om-Sed featured nighttime entertainment of its own: "Pompeii," a mammoth fireworks spectacle by the Pain Pyrotechnic Company of Chicago, "Sole Pyrotechnists to the World's Columbian Exposition." The agricultural society helped to book "Pompeii" and even contributed to paying for it, in the hope that the reenactment of the eruption of Mt. Vesuvius would attract large crowds to Des Moines, and that many of the visitors to "Pompeii" would also attend the fair.[63]

As the fair opened its gates, the *Iowa State Register* declared optimis-tically that neither the World's Fair nor the depression would crimp the state fair's attendance, claiming that "it is a full grown institution, a permanent one, that cannot be affected in the state by any outside causes." Unfortunately, the Columbian Exposition greatly diminished the state fair's displays of corn, horses, and machinery, because many large exhibitors chose to display their crops, animals, and products in Chicago.[64] The state fair had its own attractions in 1893, but these were paltry, almost laughable compared with the Midway in Chicago and the nightly eruption of Mt. Vesuvius downtown. Along with the perennial brass band, the fair's bill included a balloon ascension, in which a dog and woman parachuted from the balloon in front of the fair's grand-stand; acrobats Delmore and Lee; and a few trotting dogs.[65]

"Pompeii" enjoyed a wildly successful run in the city, and the *Regis-ter* declared that the nightly torrent of lava would increase attendance at the fair. Instead, the fireworks extravaganza's popularity backfired on the fair, siphoning off the fair's nighttime crowds and leading many people to skip the fair altogether. As a result, the fair's receipts plum-meted in 1893, and its embarrassed secretary was forced to write doz-ens of letters of apology to exhibitors as he explained that the society was unable to pay the fair's premiums. The society's financial shortfall unleashed a storm of criticism of its decision to hold the fair in 1893, forcing John Shaffer to resign as secretary after overseeing the state fair for two decades.[66] Buried in debt by the depression and by the card-board volcano, the society's new officers had learned at least one im-portant lesson from "Pompeii." Realizing that the spectacle had cost the fair thousands of dollars in lost receipts by luring patrons away from the grounds and into the city, Shaffer's successor, P. L. Fowler of Des Moines, declared that if such a spectacle were to be staged during fair week in the future, "we want it on the fairgrounds."[67]

The End of the Road

Practically and symbolically, the Chicago World's Fair marked a watershed for midwestern state and county fairs, and its gleaming White City cast a shadow over them for years. The World's Fair set a new standard for exhibitions that smaller fairs strove to emulate but could not remotely hope to attain. Many agriculturists and journalists contended that the state fair's financial difficulties resulted not merely from the depression or the fair's inability to compete with the splendors of the White City but because the fair had failed to keep pace with a rapidly changing society. Farm folk had become more sophisticated and now craved fairs that offered something more exciting than brass bands, livestock exhibits, and gigantic vegetables. If state and county fairs were to survive, they would have to entice farm families with up-to-date amusements and exhibits.[68] The *Register* observed that the World's Fair, "like a stupendous dream," had rendered state and county fairs humdrum by comparison and suggested a deeper reason for fairs' sagging popularity in Iowa and across the nation: "It is possible that we have come to the end of the road as far as old-fashioned fairs are concerned."[69]

Fair men, showmen, agriculturists, journalists, and farmers would continue to debate the role of the fair well into the twentieth century, as the fair simultaneously incorporated a host of modern educational exhibits and a dizzying array of shows, rides, and games. The relationship between the fair's educational exhibits and its amusements remained unsettled, and those who managed and attended the fair continued to debate, sometimes heatedly, precisely where the boundary ought to be drawn. As they had for decades, Iowans continued to look to the fair as a measure of their state's attainments, as a vehicle for promoting agricultural and economic development, and as a holiday from the labors of farm life. The fair's difficulties in the 1890s had forced many Iowans to question their state's future prospects. Despite the state's rapid growth in the nineteenth century, as a new century began many Iowans and midwesterners worried that their region was no longer making steady progress but was in steep decline.[70]

3 A Finer Rural Civilization

In the summer of 1906, Minnie Miller of Harlan struck up a conversation with a carnival barker at the Shelby County Fair. The carnie sweet-talked the young girl, telling her she was beautiful and vivacious enough to become a successful actress. Many stars of the stage, he assured her, began their ascent up the show business ladder by performing beneath the canvas of a carnival tent. That night, Minnie sneaked out of bed and ran away from home to join the carnival. After two weeks of searching frantically for her daughter, Minnie's mother and police officers found her performing in a sideshow at the state fairgrounds in Des Moines. "PRETTY YOUNG GIRL DAZZLED BY FAIR," ran the headline in *The Des Moines Register*. According to the newspaper, the "innocent girl" was only too relieved to be rescued from the traveling show and return to her hometown again. "It was not what I thought stage life was," she cried to the reporter. "There was no pleasure in it at all. I don't want to be an actress any more. I'll live in Harlan all the rest of my life." Minnie's brief, unhappy showbiz career offered a cautionary tale about the wiles of carnival men, the phony allure of life beyond the familiar confines of one's hometown, and the glittering amusements that threatened to ensnare unsuspecting youth.[1]

While Minnie vowed to return home, many of Iowa's young people seemed headed in the opposite direction, leaving farms and small towns for the city. At the outset of the twentieth century Americans confronted the distressing revelation that many rural people, especially women and children, felt deeply discontent with farm life and yearned for the amenities and sociability of the city. Statistics confirmed an exodus of rural Americans from farms to cities and towns, and the 1920 Census revealed that, for the first time in the nation's history, city dwellers outnumbered farmers, a revelation that occasioned hand-wringing about the demise of the Jeffersonian vision of a republic of virtuous, independent farmers. If the wellspring of American liberty lay in the countryside, it was drying up in a hurry.

Because women and children felt the keenest dissatisfaction with the isolation and rigors of farm life, the viability of the family farm was endangered not only by the appeal of town life but from within. "The 'drift to the cities,'" Iowa author Herbert Quick observed in 1913, "has been largely a woman movement," because unhappy, lonely, over-worked farm women could not be expected to remain on the farm or to raise children who would want to remain there. One of America's most urgent national problems, Quick concluded, stemmed from the discontent of the nation's farm women.[2]

Women and children had some cause to be dissatisfied with farm life, which could be isolated and lonely, especially in the winter and early spring. Women and children performed countless hours of unpaid labor on farms across the nation. The men who typically ran farms and families preferred to spend the family's income on "productive" resources, such as plows and tractors, rather than on "luxuries" for the home, such as a modern stove. As electrification, indoor plumbing, and other conveniences became commonplace in cities and towns, many Iowa farms lacked these modern amenities.[3]

But the exodus from the farms did not occur solely because women yearned for indoor plumbing or because children were bored to distraction, but as a result of powerful economic and demographic trends. Throughout American history, farm families with more than one child frequently had to decide whether to leave the family farm to one of their children, forcing his or her siblings to move elsewhere, or whether to divide the land into separate parcels, each of which might be too small to support a family. In the frontier era, when land remained relatively plentiful, young people could leave their family farm to seek land elsewhere, but as unoccupied land became scarce, their prospects for acquiring farms diminished. The increasing efficiency and mechanization of agriculture, which scientific agriculturists had avidly encouraged, accelerated the consolidation of farms into larger, more profitable units. Simply put, a growing number of people could not live on a shrinking number of farms. The growth of cities and factories beckoned many farm youths, accelerating the ominous "drift to the cities."

As historian Jackson Lears has observed, Americans have often overdrawn the contrast between urban and rural life. "City and country offered mother lodes of metaphor," he writes, which Americans used to make sense of the economic changes transforming their society in the late nineteenth and early twentieth centuries. But to pretend that city and country were unconnected, or even opposites of one another, obscures the undeniable truth: rural Americans sold their product to urban markets and purchased goods manufactured in the nation's cities, while city dwellers depended on farmers for food and as consumers. Despite some Americans' fear that "the city" was somehow undermining or corrupting "the country," the fates of rural and urban America were inseparable.[4]

While many Americans hailed industrialization and urbanization as signs of progress, others did not consider the exodus from the farm to town inevitable, much less desirable. Instead, they sought to understand why some rural Americans found farm life unappealing and to reduce farmers' physical and cultural isolation in order to slow or stop the exodus to the cities. Because women and children were thought to be especially unhappy on the farm, the federal government launched an effort to preserve rural life by making it more appealing to them. This attempt to improve rural life was both tangible and intangible, and it included ideas for improving women's and children's daily lives coupled with a campaign to promote the virtues of the countryside. To arrest the drift to the cities, the government simultaneously sought to provide farmers the amenities of town life, while insisting that farm life remained superior to it.

State and county fairs were indispensable forums for any effort to improve rural life. Fairs had for decades been considered barometers of the economic and cultural vitality of their state or county and vehicles for improving rural life. Fair exhibits could improve the conditions of farm life and encourage women and young people to remain on the farm. Ironically, though, fairs were also frequently blamed for hastening the exodus to the city. Farm families eagerly anticipated the annual fair as an occasion to travel into town to enjoy the stores, entertainment, and sociability that made city life appealing. As a writer for *Scribner's* observed astutely in 1914, "The country boy's strange unrest and passion to leave home, as yet unanalyzed and unarrested by sociology, is abetted by this annual glimpse of the world outside."[5]

Many midwesterners viewed fairs both as indicators of the condition of rural life and as important agencies for improving it. In the early twentieth century, the perennial debate over fairs' role bespoke the larger economic and demographic problems confronting midwestern farmers. Iowans fretted about the fair's "drift" toward entertainments, just as they worried about the ominous "drift to the cities."

They suspected, reasonably, that the same forces of urbanization and consumerism transforming the fair from an agricultural exhibition into a carnival were responsible for the plight of the countryside, and they struggled to preserve both the fair and rural life from the perils of urbanization.

Critics of entertainments commonly described the agricultural fair as a rudderless ship, adrift on some unseen current, and in imminent danger of running aground on the jagged shoals the show business. At the annual convention of county fair secretaries in 1913, one fair man warned his counterparts that "we are drifting away from the original ideal of a fair. In other words, we are making it more of an amusement proposition rather than carrying out the idea of old." Four years later, county fair secretaries pondered the question "Are Fairs Drifting to Entertainments?" Typically, Iowans viewed changes in the fair a bellwether of broader changes in rural life. As the *Homestead* wrote in 1921, not only the nation's population, but also the "attention of the people" had "drifted" away from rural life and from "local amusements and community affairs" toward cities. State and county fairs were significant both practically and symbolically to reformers who sought to regain control over their own society and to arrest its "drift" away from rural life.[6]

There Is No More Important Person . . . Than the Farmer's Wife

Many Americans considered the shrinking, dissatisfied rural population a matter of national urgency, and the federal government took a keen interest in the plight of the nation's farmers. In 1907 President Theodore Roosevelt appointed a commission to study farm life and offer suggestions for improving it, and two years later the Country Life Commission issued its influential report on the state of rural America. Roosevelt declared when appointing the commission that the nation's most urgent task lay in preparing farm children for farm life, and doing "whatever will brighten home life in the country and make it richer and more attractive for the mothers, wives and daughters of farmers." "There is no more important person," he wrote, "measured in influence upon the life of the nation, than the farmer's wife, no more important home than the farm country home."[7] The Commission expressed some concern about the farmer but was much more worried about his children, and especially about his wife. Its report observed that the burden of farm work, privation, and isolation fell most heavily on women, whose household tasks recurred "regardless of season, weather, planting, harvesting, social demands, or any other factor." The acquisition of household appliances and utilities widely available to their middle-class urban counterparts would alleviate some of the drudgery of farm life. These modern amenities, coupled with more efficient household management, would render farm women more content with rural life and create a home environment more likely to keep the next generation of rural Americans on the farm.[8]

No effort to improve rural life could overlook the role of state and county fairs, which had been invented for precisely that purpose. In 1911 the Department of Agriculture urged that fairs "be redirected and enlarged . . . to take advantage of the opportunities for aiding rural betterment." Within a few years, state agricultural colleges, especially their agricultural and home economics extension services and 4-H clubs, transformed and revitalized the fair's role as an agricultural and educational institution by introducing dozens of new exhibits designed to improve living conditions, nutrition, and health of farm families.[9]

State and county fairs had been created to increase agricultural productivity, which would in turn create a prosperous, stable countryside in the Midwest. In the nineteenth century, agriculturists had sought to build Iowa's agricultural economy. But by the end of the century, Americans were confronting the realization that increased productivity alone would not inevitably improve the lives of farm families, especially the lives of women and children. Instead of concentrating solely on production, government officials and university scientists now focused their attention not on building the rural economy but on preserving it, by teaching farm families how to utilize their resources more efficiently and seeking to narrow the gap, real and imagined, between country life and town life. The fair's displays of livestock and crops continued to extol agricultural bounty but now devoted attention to educating children about raising livestock and crops. Exhibits of home economics encouraged farm women and girls to adopt new standards of nutrition, home design, and cleanliness, and to become informed consumers of store-bought foods, clothing, and other products.

University-trained agricultural scientists and home economists sought to revive the fair's role as an educational institution, and the state's agricultural press never sounded more committed to the idea of the fair's importance as an educator. "Without question," the *Homestead* declared in 1907, the state fair "is the greatest educational institution in the state." Because few farmers attended college, or even high school, the fair served as "the farmer's university," teaching practical lessons to thousands of patrons for the modest tuition of fifty cents. "If properly conducted," the *Homestead* contended, "the state fair fulfills the function of a university. It answers the needs of the masses instead of catering to a fractional percentage as do our regular educational institutions." Always eager to burnish the fair's prestige and defend themselves against the accusation that the fair had forsaken its educational mission and become a carnival, the fair's organizers readily echoed the agricultural journalists and home economists who proclaimed the importance of the fair's new educational exhibits.[10] The debate over education and entertainment at the fair took on a new significance in the early twentieth century because many proponents of preserving rural life believed that the lack of diversions in the countryside ranked among the chief causes of women's and children's dislike for farm life, while others continued to inveigh against the fair's abandonment of its original commitment to agricultural education. Debate over the fair's proper role assumed a new shape, as fair men, government officials, and home economists attempted to use the fair to preserve a viable rural culture in the Midwest.

In Iowa as She Was in Eden

Exhibits created by and for women were not new at the fair, which had always offered premiums for traditional feminine skills, such as cooking, sewing, and other domestic crafts. In the fair's earliest years, exhibits of food, clothing, and other household items acknowledged women's invaluable contribution to the state's prosperity and their family's welfare. Winners in the fair's agricultural exhibits received far larger prizes, but the premium list also rewarded women's production of foods, clothing, and other household items, which contributed enormously to their families' well-being. After surveying the clothing displays at the 1859 fair, the *Northwest Farmer* hailed the frugality and resourcefulness of farm women, who sewed clothing for their families and resisted the temptation to purchase ready-to-wear clothes manufactured in other states.[11]

Women's labor not only had practical and economic value but was credited with fostering culture and refinement in the Midwest. Clothing and other household products were classified alongside fine arts in the final categories of the fair's premium list, suggesting that agricultural bounty laid the foundation for cultivation of a higher sort. In his opening address at the 1877 fair, the society's president explained that the fair's agricultural contests displayed the raw material, while "the house-keeper's department exhibits the same in more refined and concentrated form. The stronger arm of man furnishes the supplies, which have need of woman's hand to fit them for enjoyment." As the fair's exhibits grew more extensive in the 1870s and 1880s, new categories of ladies' "fancy work" proliferated in the fair's premium list, supplying tangible evidence that agricultural prosperity in turn produced refinement.[12]

Women helped to refine both life in Iowa and the fair in other ways as well. When the agricultural society purchased its permanent fairgrounds in 1885, the Iowa Woman's Suffrage Association and Women's Christian Temperance Union became the first organizations to construct buildings on the grounds. Secretary John Shaffer, keen to enhance the fair's respectability, declared that these organizations would promote "a higher standard of morality and goodness at our annual exhibitions." In 1891, women's rights advocates persuaded the agricultural society to designate an official day for women, called Woman's Day, at the fair, featuring an address by Iowa suffragist Carrie Chapman Catt. In subsequent years, suffrage and temperance advocates persuaded the society to devote one day of the fair to women's activities, and Woman's Day became a regular feature of fair week.[13]

When the state government assumed responsibility for overseeing the fair in 1900, it sought to make the fair into a more substantial and reputable institution, appropriating money to improve the fairgrounds and launching a series of construction projects over the next three decades that transformed the fair's facilities. When the fair opened in 1902, *The Des Moines Register and Leader* proclaimed it "A Twentieth Century State Fair," one at which a well-appointed fairgrounds and permanent exhibition buildings had replaced ramshackle buildings and tents. According to the paper, the improvements in the fairgrounds and in its exhibits marked the beginning of "a new era" for the fair, which would only grow more instructive, entertaining, and impressive in the future. Beginning with the construction of the fair's Livestock Pavilion in 1902, legislators appropriated funds for several permanent exhibition buildings, including a grandstand, Agricultural Building, Administration Building, Machinery Building, and livestock barns.

The fair secretary's effort to make the fair more appealing to women contributed to this transformation of the fairgrounds. In 1903 the Board of Agriculture remodeled the fair's Horticultural Hall to create the Women's Building. He predicted that this new facility would "great-ly popularize our fair" by making the exposition more attractive to women, especially those with young children. Rather than a meeting place for women's organizations, the new Women's Building primarily offered a refuge from the heat and fatigue that frequently sapped fair-goers' energy. As the fair's secretary pointedly informed suffragists, temperance advocates, and members of other women's organizations, "this building is not to be a meeting place for women's clubs, but a rest-ing place for women and children."[14]

In 1910 the Fair Board selected the Chicago firm of landscape archi-tect O. C. Simonds to devise a formal plan for the grounds. Influenced by the goals of the Country Life movement, Simonds believed that a properly designed fairgrounds, like an attractive farmstead, would help preserve rural life. As Simonds later wrote in his influential book *Land-scape-Gardening*, properly landscaped farms and parks could bridge the gulf between urban and rural life and "bring about a friendly rela-tion between the people of the city and those of the country." Simonds's plan for the fairgrounds retained much of the basic layout created in the 1890s, enhanced by the proposed addition of several impressive neo-classical buildings, which would lend the grounds the appearance of a university campus. Simonds eliminated the hodgepodge of wood-frame structures built by implement manufacturers, replacing them with the enormous steel-and-masonry Machinery Building. He also plotted the fair's campground in an effort to make camping at the fair more conve-nient and to encourage more families to attend the fair and to lengthen their stay. Although Simonds's most ambitious ideas never made it off the drawing board, the state government did appropriate funds for the construction of several substantial buildings on the grounds over the next two decades.[15] As the fair's managers pored over blueprints and worked to improve the fairgrounds, women's organizations, especially the Iowa Congress of Mothers, urged them to add a building devoted to women's activities, and in 1913 the legislature appropriated $75,000 to construct the fair's Women's and Children's Building. This building, the Fair Board promised, would make the fair much more attractive to women and children.[16] The building's planners and the fair's secretary envisioned the new building housing exhibits related to homemaking and, especially, childrearing. Arletta Clarke, wife of Governor George W. Clarke, declared at the building's dedication ceremony that the new structure was consecrated to the belief that "whatever activities wom-en might undertake, the uppermost purpose of their lives would be the welfare of childhood." A writer for the *Homestead* summed up the building's implicit message after viewing the Women's and Children's exhibits a few years later, observing that the building and its exhibits embodied the belief that women's interests could not be separated from those of the farm family: "So much for the women's own building. But after all, friends, isn't one of the finest things of farm life the fact that there are no lines of division in the family's interest?"[17]

To ensure that the new building would remain devoted to homemaking and childrearing, the Fair Board entrusted oversight of women's and children's activities at the fair to the Iowa Congress of Mothers and Iowa Federation of Women's Clubs, decreeing that the Iowa Women's Suffrage Association and other overtly political organizations would not be allotted space in it. Although suffragists did secure the state's ratification of the Nineteenth Amendment in 1919, they never gained admission to the Women's Building, which remained, at least ostensibly, free from politics and exclusively devoted to women's domestic roles as homemakers and mothers.[18]

Iowa's Greatest Product

The fair's exhibits had always hailed agricultural improvement as the benchmark for measuring civilization's progress. Barns filled with immaculately groomed purebred livestock and farm displays bursting with a cornucopia of fruits and vegetables attested to agricultural prosperity, which would invariably contribute to progress in manufacturing, domestic crafts, and the arts. Some midwesterners decided that the same techniques used to produce and judge livestock and crops could be applied to breeding, improving, raising, and measuring human beings. Beginning in the 1910s, the most popular event in the Women's and Children's Building was scientific baby-judging, in which Iowans strove to improve human beings, just as they had improved the quality of the state's herds of cattle and hogs.

As early as 1872, one farmer wrote to the *Iowa Homestead and Western Farm Journal* to suggest that state and county fairs offer premiums "for excellencies and eccentricities of *human* as well as *animal* and *vegetable* nature [emphasis in the original]." In 1876 Cedar Rapids merchants sponsored a "baby fair," offering prizes for the baby "with the strongest lungs," "the fattest boy babe," the girl "who has the best shaped head, and the best head of hair," "the best muscled pair of twin babies," and other categories. Throughout the late nineteenth century, the fair frequently included a baby beauty contest. Comparisons between livestock and human beings sometimes proved flattering to the animals. After observing the swine exhibit at the 1877 fair, one reporter remarked flippantly that it "almost makes one a convert to the Darwinian theory. There are pigs as beautiful as babies, and hogs as neat as horses. The hoggish countenance has been subdued to something almost human, and the repulsive snout and form have been transformed into something bordering very nearly on the beautiful."[19]

Some Iowans took seriously the idea that farmers ought to devote at least as much effort to producing better human beings as they did to breeding cattle and hogs. In 1883 agricultural society president William Smith pointed out that while farmers lavished attention and expense on breeding improved stock, too many young people chose their mates with less care than they would exercise when buying a horse or a pig. The result: offspring as dim as "a lamp lighted which contains no oil." Smith urged farmers to apply their experience in raising purebred livestock to the rearing of their own offspring and predicted that Iowa's fertile land and agricultural bounty would enable the state to produce superior human beings as well. "The country which produces the finest domestic animals," he observed, "should, under proper influences, produce the finest specimen of manhood; and while caring for our herds we should never lose sight of our children." The ambiguity of Smith's remark was telling: on the one hand, it expressed the commonplace belief that agricultural productivity would ultimately contribute to success in other endeavors; on the other, it betrayed a nagging suspicion that Iowa's remarkable economic progress had not improved the quality of its people, who lagged behind like so many scrub cattle.[20]

Coincidentally, at the same moment Smith was urging Iowans to breed better offspring, British naturalist Francis Galton, cousin of Charles Darwin, coined the term "eugenics," advocating social policies calculated to produce a healthier, more intelligent populace by encouraging those deemed healthy and intelligent to reproduce. Eugenicists soon distinguished between "positive" eugenics, which sought to encourage reproduction by healthy, intelligent, and otherwise "fit" individuals, and "negative" eugenics, which sought to discourage or even prevent supposedly "unfit" people from bearing children. Farmers had bred animals for centuries and had become much more systematic about breeding since the eighteenth century. Eugenicists consciously sought to breed an improved human race, and contended that regulating reproduction could benefit humanity by weeding out "unfit" parents and encouraging reproduction by those people deemed intelligent and healthy enough to produce offspring.[21]

While attending the Audubon County Fair in 1911, Mary T. Watts of Audubon marveled at "the wonderful improvement that had really been made in the condition of livestock" but observed the alarming disparity between the magnificent animals in the fair's barns and the somewhat less impressive human specimens strolling the fairgrounds. Watts decided that fairs should undertake a new mission: to improve what she termed the state's "greatest product"—its babies.[22] At Watts's suggestion, the state fair inaugurated a baby health contest in 1911 under the auspices of the Iowa Congress of Mothers (prior to 1911, the organization had recommended that women not expose their young children to the fair's unhealthy environment!). Billed as "a strictly scientific affair," the new contest replaced its decidedly unscientific predecessor, the baby beauty pageant conducted by the Central Church of Christ. Judges rated the babies on a scorecard designed by Dr. Margaret Clark of Waterloo, who, according to Watts, "used a live stock scoring card as a model, making, of course, necessary changes to meet the requirements for the human body." In order to emphasize that the new contest was not merely a beauty pageant but a serious, scientific endeavor, the judges conducted their examinations behind closed doors, barring the public from observing the contest or even seeing its eventual winners.[23]

Despite the veil of secrecy shrouding the competition, baby judging generated tremendous enthusiasm, prompting the Fair Board to appropriate $500 for the competition and to make it an official part of the fair's schedule in 1912. The board and the contest's organizers realized that they could attract a larger audience and publicize scientific childrearing more effectively if fairgoers were permitted to witness the exhibit, and so an auditorium was constructed with a glass-walled judging room so spectators could watch as judges examined the babies. Physicians and experts on child development also offered lectures and exhibits on various aspects of "scientific" childrearing to the audience.[24]

Baby judging immediately became the fair's most popular attraction, and hundreds of people waited patiently in line for their turn to catch a glimpse of the judging. The extraordinary appeal of the baby health contest is not altogether surprising, since babies are marginally more interesting to look at than livestock or vegetables. Eager to learn about new scientific ideas concerning nutrition and childrearing, women packed the lectures by physicians and psychologists. The contest was not only popular with spectators but attracted many entrants as well: the first contest in 1911 drew fifty contestants; within five years the number had swollen to five hundred, and by the early 1920s fair officials were forced to limit the number of entries, which consistently exceeded seven hundred. The contest's winners became infant celebrities. The governor himself often awarded the grand prizes to the contest's winners, whose photographs were published on page one of the state's newspapers. Fittingly, these blue-ribbon babies assumed their rightful place at the head of the fair's "Grand Million Dollar Live Stock Parade" before the audience in the fair's grandstand. Dubbed the "Iowa idea," or "Iowa experiment," scientific baby judging spread rapidly throughout the United States and Canada. In 1912, delegates to the American Medical Association convention in Atlantic City endorsed scientific baby health contests. Within three years of inventing Iowa's health contest, Watts, who helped organize contests across the nation, declared that baby judging "has come to be a great movement."[25]

As with the fair's livestock contests decades earlier, scientists insisted on securing trained, expert judges to evaluate babies. After observing the 1913 contest, Professor Charles Seashore, a psychologist at the University of Iowa, stated that physicians lacked the specific expertise to score babies and announced his plan to develop a graduate-level course in baby judging at the university, foreseeing "a need in future years for expert judges who have had the advantage of laboratory work in the university."[26] Inside the special auditorium at the fairgrounds, spectators and anxious parents watched as physicians scrutinized infants from head to toe, dutifully recording even the slightest imperfection on their scorecards. Parents unaccustomed to this precise, quantitative measure of their baby's health were often disheartened to learn that "the perfect darling may have ears a trifle too large, or a chin a little out of line or a stomach a quarter inch too big." Judges sized up the parents as well. To enter their child in the contest, parents were required to complete an extensive questionnaire concerning their ancestry, occupation, use of alcohol and tobacco, and childrearing practices.[27]

In order to ascertain that competitors possessed sound minds, as well as sound bodies, the judges administered a mental test, in which they carefully evaluated a youngster's facility at playing with wooden blocks and a rubber ball, observing a moving toy monkey, and making marks with a pencil. As the *Register* flippantly noted, the data obtained from these tests was "vitally important to the psychologist, nonunderstandable to the worried mammas and just a joke to the babies." The successful competitor had to be bright, but not too bright: psycholo-

gists penalized precocious babies for being overeager. A touch of me-
diocrity was considered a virtue, since proponents of scientific childrearing encouraged parents to raise children who would fit easily into
society, rather than geniuses.[28]

By the mid-1920s, the contest's organizers began to assess not only
the child's intelligence but his or her personality or "mental and nervous hygiene" in an attempt to diagnose potentially antisocial tendencies. As in the intelligence tests, successful contestants were judged to
be those in the middle of the pack, neither too dull nor too bright. Judges penalized children who responded sluggishly when offered a toy for
"apathy," but also marked down those who responded too eagerly for
"grabby" behavior. According to one of the psychologists who administered these tests, "A brazen child is marked off as many points as one
who is too shy. We would rather have a child move slowly and be more
natural than be over sophisticated. Sophistication is a danger signal."[29]

The baby health contest ostensibly aimed to determine the healthiest childrearing practices and to publicize these practices to fairgoers,
and the contest and its accompanying exhibits and lectures emphasized that weak or sickly children could become more robust with proper diet and care. But the ugly specter of negative eugenics—of using the
contest not only to promote better childrearing but to stigmatize, discourage, or even prevent reproduction by those deemed mentally and
physically unfit—lurked behind the wildly popular spectacle of baby
judging. Mary Watts acknowledged in 1914 that it was virtually "impossible to control the ancestry of our children as rigidly as we control the
breeding of live stock," but she advocated "the sterilization of the unfit"
in order to eliminate "defectives," such as "the criminal, feeble minded
and insane."[30]

While doctors and scientists disagreed over the relative importance
of heredity and environment in producing healthy babies, they overwhelmingly agreed that babies raised in the countryside would prove
more robust than their enervated urban counterparts. To virtually everyone's dismay, during the first few years of the contest urban babies
consistently outscored rural babies. Baby judges attributed this star-
tling result to rural parents' persistence in clinging to outmoded folk
wisdom about childrearing, raising "old-fashioned" babies in closed
rooms and treating their illnesses with home remedies instead of avail-
ing themselves of modern medical science. A writer for the *Homestead*,
though, disagreed, contending that urban babies had prevailed because of the prejudices of the contest's judges. Unlike most of the fair's

other competitions, "the baby contests have been conducted almost entirely by city-bred folks" with new-fangled ideas about childrearing. Unwilling to subject their children to the skewed opinions of doctors and judges from the city, country people were reluctant to enter their children in the contest. The inclusion of a few farm women among the judges, the *Homestead* declared, would result in a contest filled with farm babies who "will make these lace-frilled, prize-winning darlings from the cities look like invalids when it comes to good health." For most Iowans, however, the unsettling revelation that rural children lagged behind their citified peers underscored the urgency of the baby health contest as a means to improve the well-being of rural Iowans and arrest the decline of the countryside.[31]

Worries about the viability of rural life led many spectators to scrutinize the baby contest for reassurance that the countryside remained a superior environment in which to live and raise children. While the baby judges sometimes warned that too many farm children had not been blessed with especially good heredity or had been raised improperly, the contest became the centerpiece of a concerted effort to improve the state's "greatest product" and to improve the health and happiness of farm families. Scientific baby judging also helped bolster the fair's role as an educational exhibition, and the contest provided a perfect antidote to criticism of the fair's burgeoning array of Midway rides, games, and spectacles in the early twentieth century. As *Wallaces' Farmer* remarked, many of the fair's attractions provoked "a sense of repulsion which needs to blot it out the wholesome sight of a man bringing his two-year old baby to the baby health conference and caring for it through the various tests, measurements and examinations." The wildly popular baby contest became central to the effort not only to improve the health of Iowa's children but to preserve the rural life generally and to burnish the fair's image as well.[32]

The Flower of Farm Youth

A few years after the state fair began to determine the state's healthiest babies, physicians created a similar contest for the state's teenage 4-H members. As with scientific baby judging, Iowans pioneered this health competition for adolescents, which quickly spread to fairs throughout the country. In 1919 Dr. Caroline Hedger of Dubuque lamented that the fair awarded "a blue ribbon for the steer, but none for the lad" and instituted a contest to determine the state's healthiest teenage boy and girl.

Like the baby health contest, the effort to determine Iowa's healthiest teenagers promoted eugenics and publicized the superiority of farm life. (This competition, however, unlike the baby contest, did not pit contestants from town and country against one another and was restricted to 4-H club members, most of whom resided on farms.)[33]

To determine the state's healthiest youth, judges examined contestants in microscopic detail. The judges' "report card" of Alberta Hoppe, the 1926 girls' champion, conveys the judges' exactitude in scrutinizing her body for the slightest imperfection: "Doctors in the final examination of Alberta scored off .03 for irregular teeth and for malocclusion; .1 for abnormal eyelids; .2 for enlarged glands; .2 for slight irregularities of form; .2 for antero-posterior curvature of the spine; .2 for irregularity of her feet and legs and .1 for her gait." Even the fair's blue-ribbon livestock were not subjected to scrutiny this obsessive.[34] In addition to crowning overall champions, the judges also awarded prizes to the girls and boys with the most perfect teeth, feet, and posture. The unrequited quest for Iowa's own Cinderella—that is, for a young woman with perfect feet—became a virtual fetish in the 1920s. The director of girls' 4-H clubs in the state rhapsodized at the prospect that "the perfect pedal extremity this year, if located, will achieve more fame than the fairest face."[35]

The 4-H health contests measured more than physical health. Judges and journalists alike hailed health champions not only for their physical attributes but especially for the ostensibly rural virtues that their robust health manifested. Judges of the 1926 competition were relieved to find "no 'string bean flapper types among the Iowa farm girls in the contest" and "expressed their approval of the plump, well-rounded bodies of the farm maids." (To allay potential misgivings about dozens of girls being subjected to public physical examinations, *Wallaces' Farmer* reassured its readers that "the girls wear neat blue health suits throughout their examination.") The 1926 winner, Alberta Hoppe, whatever her "slight irregularities of form," was a paragon of moral virtue, who resisted the temptations of fashion, makeup, and entertainments. The *Register* noted approvingly that "she has never worn a corset or a high-heeled shoe. She uses no powder or rouge, cares nothing for boys and dates, does not dance, and rarely goes to movies." The *Register*'s headlines about 1937 champions Estella Vermeer and Junior Clayton proclaimed their ordinariness and self-restraint in headlines that read like midwestern haiku:[36]

Dislikes Fancy Clothes—
No Cosmetics for Champion
Estella Leads
A Normal Life

Regular Boy
Who Doesn't
Often Eat Pie

Male champions, such as 1929 winner Kenneth Redfern of Yarmouth, typically felt "a natural antipathy for an excess of pastry, for liquor or tobacco," shunned movies and dating, and acquired their strapping physiques from years of doing farm chores. Most important, 4-H champions unanimously intended to remain on the farm. "I intend to follow up farming. I like it," Dudley Conner of Malvern proudly informed the *Register* upon winning the 1933 contest. "The City? I should say not. I love the farm and I always will," declared 1936 girls' champion Edith Belknap. These winners' determination to remain on the farm suggested that the exodus of rural youths to the city could be arrested, and that the countryside would not be peopled solely by dull plodders too inert to pick up and move to town. Like the baby judging, the 4-H health contest aimed to preserve the family farm not only by improving the health of farm youths but by promoting a more flattering image of rural life.[37]

The Fair Has a New Meaning

In order to make farm life more appealing to young people, the Country Life Commission urged the creation of a federally funded rural extension service, which would enable university-trained extension agents to disseminate knowledge about scientific agriculture and home economics to farm families, especially to young people. In 1914 the federal government adopted the commission's recommendation by establishing the 4-H (the four Hs stood for "head, heart, hands, and health") under the provisions of the Smith-Lever Act. American entry into World War I temporarily diverted the federal government's attention and resources from efforts to improve rural life, but the 4-H spread across the nation in the 1920s. Local 4-H clubs brought farm girls together to study home economics, while farm boys concentrated on agriculture, all under the supervision of university extension agents. 4-H leaders sought not only to educate rural youths about home economics and agriculture but to provide them with a social organization that would diminish the isolation of farm life. The 4-H soon played a large role in the lives of many rural youths and fundamentally transformed many of the agricultural exhibits at state and county fairs.[38]

Despite some worry that the fairgrounds was not an altogether wholesome atmosphere, young people had always attended the fair and had occasionally contributed to its exhibits. Before tractors relegated horse-drawn plows to displays of bygone farm implements, the fair's plowing match, in which boys competed to turn the most ground and plow the straightest furrows, attracted enthusiastic crowds. In the 1870s, the fair added a separate department for young people, and "school exhibits," displaying the achievements of the state's public school students, endured well into the twentieth century. Still, most of the fair's exhibits and competitions in the nineteenth century were designed to appeal primarily to the male farmer. In an effort to increase attendance, advertisements for the fair frequently urged farmers not to leave their wives and children home during fair week, reminding them that the fair also offered activities of interest to women and children.[39]

The creation of exhibits saluting the accomplishments of 4-H members in the 1920s gave young people a much larger role in the fair and transformed agricultural displays and competitions at state and county fairs throughout the United States. These new exhibits also revived fairs' reputation as educational institutions and contributed to the effort to preserve rural life. Formerly an opportunity for farmers to exhibit their products, fairs now became important venues for educating young people and persuading them that farm life remained satisfying and respectable. 4-H clubs, exhibits, and competitions taught young people that farming was not slow-paced drudgery but was an indispensable job that required considerable knowledge and skill.

Since the fair's inception, its livestock judging had been a competition between adult farmers and breeders, but 4-H contests allowed young people to compete to raise the finest hog, cow, or sheep. The number of young people showing livestock grew so rapidly in the early 1920s that the fair had to limit the number of 4-H exhibitors by using county fairs to determine which animals were worthy to compete at the state fair.[40] 4-H displays of cooking and domestic handicrafts educated girls in scientific home management in an effort to make farm women's work more appealing. 4-H programs reinforced the gendered distribution of labor on the farm: boys learned about crops, machinery, and livestock, while girls concentrated on cooking, sewing, and other domestic skills. Girls sometimes raised animals and competed in 4-H livestock contests, but boys did not sew clothes or can preserves. 4-H exhibits, unlike the fair's previous agricultural displays, focused not only on the final product but sought to teach fairgoers about the process of raising or creating it. 4-H members exhibited animals, crops, handicrafts, and foods in the hope of winning a ribbon and a premium, but they also participated in a range of "demonstration" exhibits in which they gave public presentations about new techniques of agricultural science and home economics. Here? American agricultural fairs initially had been predicated on the belief that most people learned by looking at "the thing itself, in actual view," as one nineteenth-century agriculturist put it. Fairgoers studied prize-winning animals, apples, or quilts and strove to emulate these outstanding examples when they returned home. The fair's new exhibits, organized by university-trained agricultural extension workers, espoused a different theory of education, in which fairgoers listening as extension workers or 4-H members explained how to fatten an Angus cow for market or can green beans. Extension agents not only taught 4-H members about agriculture or home economics but trained them to convey their knowledge to others. Many 4-H competitions rated the effectiveness of "demonstration teams," as judges scored contestants' ability to explain some aspect of scientific agriculture or home economics to an audience rather than on their actual product. These competitions sought not only to disseminate knowledge about agricultural science and home economics to the audience but to train youths who embodied the virtues of rural life and refuted the stereotype of farm people as ignorant and resistant to new ideas.[41]

The state's 4-H leaders, affiliated with the Extension Service at Iowa State College, distributed to local chapters detailed suggestions for fair demonstrations, which scripted contestants' words and actions and included plans for creating signs and other visual aids for the presentations. State 4-H leaders furnished local clubs with detailed scripts for demonstration teams, and judges scored competitors according to their faithfulness in following that script. A 4-H outline for an "approved shoe demonstration," for instance, spelled out the words and actions for both members of the demonstration team to use to explain "how to be a wise consumer and purchase comfortable, durable shoes." After contestants completed their demonstrations, judges evaluated their performance: "Stand up straight"; "Girl #2 did not appear congenial"; "Be sure to lift examples high enough for all to see"; "Speak more slowly and clearly."[42]

In the 1920s, the 4-H became the mainstay of the fair's agricultural competitions and home economics exhibits. Before 1920, 4-H exhibits comprised only a few displays of home canning techniques, and the 1920 4-H girls' exhibit included only fifty girls from fifteen counties. During the 1920s, Josephine Arnquist of Iowa State College, who oversaw girls' 4-H projects in the state, expanded the organization's fair exhibits, adding displays on food preparation, clothing, and home decorating. Arnquist urged county 4-H leaders to send more exhibits and representatives to the fair in order to prove the organization's value to the state Fair Board, and the number of 4-H members participating in the fair grew so rapidly that the state's 4-H director, Paul C. Taff, worried "that the facilities of the State Fair cannot keep up with it." By the mid-1920s, more than twenty thousand Iowa girls were 4-H members, and four hundred 4-H girls participated in the annual fair, where they resided in a dormitory constructed to house them. Fair week was no idle vacation: even dormitory life was regimented and competitive, as girls vied to win awards for adhering to dormitory rules, keeping their uniforms and quarters clean, maintaining exhibits, and arriving punctually for meetings and competitions.[43]

Proponents of 4-H exhibits also touted these exhibits as proof that the fair had not drifted away from its agricultural moorings and toward the dangerous whirlpool of amusements. The *Homestead* declared that "the fair is becoming more and more an educational event and less and less the somewhat spotty recreational stunt of years ago. Going to the fair means something different to boys and girls of the present generation than it did in years agone." According to *Wallaces' Farmer*, 4-H exhibits had rescued the fair from becoming a carnival of pleasure seeking. "Merry-go-rounds and ice cream cones are now incidental," the magazine claimed, "while the important business of the fair for the boys and girls is getting the calf or the pig to look well in the showring, or learning the parts to be taken in a club demonstration team. The fair has a new meaning to the boys and girls, and the boys and girls have given the fair a new significance." By transforming the fair, the magazine claimed, the 4-H might ultimately transform the entire state, ushering in "a finer, greater rural civilization in Iowa."[44] The fair's own newspaper, *Greater Iowa*, declared that "it is because of ideas received at the State Fair and because of taking part in various State Fair enterprises that boys and girls from rural homes have been able to recognize that their best interests were in their present environments." If so, the

4-H had attained its original and most important goal: stemming the exodus of young people from the countryside to the city.[45]

The 4-H strove not only to improve the conditions and the image of farm life but to reduce the disparity, both real and perceived, between rural and town life. As a result, the 4-H simultaneously hailed the virtues of farm life and sought to acquaint farm families with the products and styles of a burgeoning consumer economy; 4-H exhibits commonly depicted farmers as resourceful, self-sufficient, and uninterested in store-bought goods and commercial entertainments. Exhibits regaled viewers with thrifty tips for stretching farm income, and 4-H members competed to make attractive furnishings from orange crates and other cast-off items and to sew dresses from feed sacks. But in order to bridge the gulf between town and country, the 4-H sought not only to encourage self-sufficiency but to acquaint farm girls with newly minted professional standards of home economics and the knowledge necessary to participate in a consumer economy. The members of 4-H learned not only to can tomatoes and sew their own blouses but to become informed consumers capable of making prudent choices when purchasing groceries, clothing, or housewares.

Especially during the Great Depression, 4-H leaders encouraged young people to resist the lures of consumer society and commercial entertainment. In 1931 the 4-H introduced a competition to reward the girl who kept the best account book of her income and expenditures. Winner Aletha Paul earned the judges' praise for her extraordinary frugality, having spent only one dollar on cosmetics over the past year. *The Des Moines Register* noted approvingly in 1932 that some Iowa farm girls "spend less than ten cents a month on ice cream and candy and don't attend a movie more than once in three months." In addition to the practical benefit of learning to budget one's money, these contests bolstered the image of wholesome farm girls, who still sewed their own dresses, saved their pennies, made their own fun, and were perfectly content with farm life.[46]

Frugal, industrious 4-H girls became the very embodiment of rural virtues and the viability of country life. During fair week in 1934, Iowans picked up their *Des Moines Sunday Register* to find a front-page article and photographs contrasting the thrifty, sober habits of Arline Kline, a 4-H girl from Waterloo, with a titillating peek at the dissolute life of Irene Tusky, a showgirl from New York who performed nightly in *World's Fair Scandals* at the Orpheum Theater. As the newspaper observed, "In Play or Work Their Worlds Are Ever Apart." Arline, her 4-H uniform neatly pressed, her hair styled plainly, and her face unadorned by makeup, did not smoke, drink, dance, or go to movies, and was already an accomplished cook and seamstress. She could be found in the pew at the Church of the Brethren every Sunday and for revival meetings during the week. Her idea of a good time consisted of attending meetings of local civic organizations.[47]

Irene, glamorous in her polka-dot dress and dark eye shadow, represented Arline's opposite in appearance and behavior. She had no clue how to fry an egg, and her hectic rehearsal and performance schedule often forced her to grab lunch on the run. She performed three to five shows daily, and "even though her spirits may be low, yet she must have that ever-present smile for theater patrons," because "there is many a girl who can and will take her place if her performances are unsatisfactory." She occasionally found time to squeeze in a Catholic mass, often after a sleepless Saturday night. Lest readers draw the wrong conclusion about which girl's life was preferable, the *Register* emphasized that Arline's life, while unexciting, offered the solaces of home, family, community, and honest labor unknown to the transient showgirl. Like the fair itself, though, the *Register*'s effort to extol the virtues of farm life simultaneously tantalized readers with its glimpse of glamour and sophistication, even as it championed the superiority of rural life.[48]

Right Living

Just as the 4-H strove to make rural life and the fair more attractive to young people, professional home economists sought to make them more appealing to women. Although women had displayed foods, clothing, and other items at the fair since its inception, many of the fair's exhibits were primarily of interest to men, and many farm women either did not attend the fair or found much of it tedious. "It used to be when a woman came to the state fair that there was nothing for her to do but follow the menfolk around all day," feigning interest in plows and tractors, recalled one woman in 1927. Far from a leisurely outing, the task of keeping one's family well fed and clean at the fair entailed "more work than play for her." In the early twentieth century, professors and extension workers at Iowa State College sought to improve farm women's lives and devised a host of new fair exhibits to interest women and instruct them in the proper techniques of "household management." As a result, attending the fair now included more activities for women, and the annual event became "just as much of a woman's fair as a man's."[49]

As early as the 1870s, home economist Mary Welch of Iowa State College, wife of college president A. S. Welch, exhibited some of the techniques and tools of modern home economics at the fair, but the college's home economists did not regularly stage exhibits at the fair until 1907. Over the next decade the federal government's efforts to improve country life, college extension agents' efforts to educate rural Iowans, and the Fair Board's interest in making the fair more attractive to women coincided, and a wide variety of home economics exhibits quickly became one of the fair's leading educational features.

When the fair opened its new Women's and Children's Building in 1914, home economists sought to make the building a clearinghouse for the latest discoveries in nutrition, childrearing, and other topics of interest to mothers and homemakers. But home economists had even larger ambitions. As Neale S. Knowles, a home economist at Iowa State College, declared, proper home management entailed far more than serving one's family a balanced diet or maintaining a clean kitchen. Home economists aspired not merely to make women into better homemakers but to rid the state of a host of social ills, which, in the language of Progressive Era reformers, could be distilled into in a single word: inefficiency. They sought to host programs in the fair's Women's and Children's Building that would contribute to "the great national effort to increase interest in 'right living,'" by reducing "disease, poverty, crime and all phases of inefficiency, to the minimum."[50]

The state fair offered home economists a unique opportunity to promote their ideas to a large audience, but it hardly enabled them to reach all of the state's farm families. The college's agricultural and home economics Extension Service disseminated knowledge about scientific agriculture and household management across Iowa by dispatching county extension agents across the state to educate and uplift farm families. The underlying premise of "extension," that the university should spread education to all corners of the state, was precisely the reverse of the fair's method of educating those farmers who congregated at the annual exhibition. The advent of agricultural and home economics extension revitalized the fair's educational exhibits but also marked the culmination of college supplanting the fair as the principal vehicle for educating the state's farm families. The college's home economists considered the fair an important annual showcase for their efforts to improve farm life across the state. Educational exhibits created by college home economists, rather than competitions for premiums, became the centerpiece of the fair's household displays in the 1920s. Advertisements for the 1922 fair announced that the homemakers' exhibit would now be "under the direct supervision of educational specialists and home department agents who have made thorough study of this work." Women still entered homemade dresses and pies to compete for premiums, as they had since the first fair in 1854, but university-trained home economists believed they could more effectively improve domestic life through lectures and displays on nutrition, textiles, childrearing, hygiene, decor, and shopping.[51]

Rather than salute the achievements of individual homemakers, home economics exhibits tallied the overall success of home economics extensions in spreading scientific household management throughout the state's ninety-nine counties. County extension workers dutifully recorded the results of their meetings with farm women on detailed scorecards distributed by the Extension Service, and the fair provided them an ideal opportunity to tally and publicize their efforts to inculcate "right living." As Neale S. Knowles put it in 1931, exhibits of extension work in the state's counties aimed not only to instruct fairgoers how to be better cooks, seamstresses, and mothers but to provide "a round-up showing the year's accomplishments" in food and nutrition, clothing, home furnishing, child development, and home management.[52]

Home economists tabulated precisely the effectiveness of extension work among farm families. The fair, which had always tallied crop yields and hogs' weight, now tallied statistics on Iowans' household habits. The march of progress could be counted precisely: in 1923, the second year in which extension workers displayed county exhibits at the fair, they proudly reported that a grand total of precisely "55,187 people have adopted suggestions pertaining to better ways of clothing the family, furnishing the home, managing their work and feeding their families and friends." Women in one township had used pedometers and "had saved 40,000 steps by 'changing their ways.'"[53] Displays of county extension projects at the 1928 fair measured extension workers' success in introducing improved cooking, cleaning, and childrearing practices. Extension workers reported that women in Dallas County had "disposed of 341 useless articles, refinished 175 floors, and removed 141 unsightly spots." Not to be outdone, women in Scott County had successfully adopted new childrearing techniques, eliminating 92 negative attitudes, 73 abnormal fears, 47 temper tantrums, and 22 thumb suckings among their children.[54]

Although home economists hoped that county exhibits would inspire fairgoers to adopt better household management in their own homes and communities, the new exhibits only mystified some viewers. Instead of tangible displays of blueberry pies, blouses, or needlepoint, these displays consisted of placards promoting the accomplishments of home economics extension. One writer for *Wallaces' Farmer* complained, "To the uninformed, the booths seem a maze of maps and charts," intelligible only to extension workers.[55] Home economics extension agents frequently charted their progress on county maps, shading in each farmstead conquered by modern domestic science. They envisioned a day when they could draw "perfect" maps for each of the state's ninety-nine counties, ultimately reaching "the last woman on the farm." Ultimately, extension workers sought to spread their influence throughout the entire state, eradicating every last semblance of behaviors they considered "outmoded" or "incorrect."[56]

Keeping Up with the Do Wells

As home economists strove to improve the conditions of farm life, they often stigmatized those benighted farm families who stubbornly refused to adopt "right living." In their effort to combat the stereotype of farmers as backward, home economists scarcely concealed their contempt for those farm families who did not adopt modern household management. As the *Homestead* observed in 1922, county extension workers were creating "a new type of country woman who is gradually and surely pushing the notion of the lonely, overworked farmer's wife into the discard," and those farm women who stubbornly clung to outmoded, unscientific housekeeping practices were no longer the objects of sympathy, but of scorn. Reporting on the inaugural exhibit of home extension projects at the 1922 fair, the *Homestead* reported that "the farm woman who does not realize her own value as a producer and homemaker does not get much sympathy from her progressive sister." A display of labor-saving ideas and gadgets sponsored by the Farm Bureau a few years later put it even more bluntly: "If a farm woman is a slave, it's her own fault."[57]

Home economists sometimes insisted that a few simple housekeeping tips could enable any family, no matter how poor, to live reasonably well. Purchasing a handful of inexpensive household utensils and adopting scientific housekeeping techniques and better management of the family finances, according to home economists, could almost magically transform poor, discontented rural residents into prosperous, happy families. The Scott County exhibit at the 1926 fair recounted the tale of the Notsogood family, "who had nothing in their life but discouragement, hard work, long hours, ill health and quarrels over family funds" until Mrs. Notsogood prudently adopted "good management." She built a white picket fence to keep the chickens out of her yard, and a lawn and flower garden suddenly sprouted around her house. Soon she acquired running water, electricity, and a car. Befitting their new station in life, the Notsogoods even changed their name to Do Well.[58]

Well-intentioned home economists offered farm families useful information on household management and achieved real improvements in health and nutrition. But home economists' effort to promote the image of happy, healthy farm families simultaneously stigmatized poor, less educated, or just plain stubborn rural families who refused to adopt scientific household management. Amid the protracted agricultural depression of the 1920s, a bit of penny-pinching and a few household tips might prove beneficial, but they were hardly sufficient to rescue many families from poverty, and, as the Depression deepened in the 1930s, many Notsogoods became Worses.

Despite their occasional tendency to underestimate the difficulty of climbing out of rural poverty, home economists were well aware of the resentments and feelings of inferiority that farm families often felt toward their better-clothed and better-housed neighbors in town. As Mrs. Ellsworth Richardson, chair of the Women's Committee of the Iowa Farm Bureau, informed the Fair Board at its annual convention in 1926, better home furnishings and more stylish clothing might not be necessities, but could eliminate one of the chief sources of dissatisfaction with farm life. So long as farm residents felt self-conscious about their appearance when they traveled into town, they would invariably consider rural life—and themselves—inferior. In order for rural people to shed the stigma of hickishness, they needed to participate in America's consumer economy and purchase the same clothing and goods available to their neighbors in town, so they could be "just as well dressed as the persons with whom they rub elbows."[59]

While home economists measurably improved the quality of life for many farm women and their families, they assumed that women would continue to fulfill their traditional gender roles of wife, mother, homemaker, and now added to these a new role as consumer. In the 1920s, and even amid the wrenching Depression of the 1930s, the fair's home economics displays, like its 4-H exhibits, paid less attention to farm women's role as a producer of clothing and food for their families and emphasized their new role as a consumer of clothing and housewares. To some extent, attending the fair had always afforded farm families an opportunity to shop for goods that they could not easily obtain at home, and merchants on the fairgrounds and store owners in Des Moines did brisk business during fair week.[60] But home economists hoped to make farm women into well-informed consumers by familiarizing them with the growing array of foodstuffs, clothing, and housewares that most Americans now purchased, rather than producing them at home. "Let's Go Shopping," a skit at the 1936 women's exhibit, showed "how a store's policies were changed through the influence of informed consumers," while lectures such as "Your Money's Worth in Clothing" acquainted farm women with "information valuable to the consumer in wading through labels and manufacturers' terms in selecting textiles and ready-to-wear." Sizing up the women's exhibit at the 1937 fair, *Wallaces' Farmer* stated that it addressed "economic problems quite as important to the farm home as parity prices or foreign markets are to the farm. Whether it be silk hosiery or canned goods, the farm woman consumer displayed her determination to get her money's worth for her purchases."[61] Proponents of home economics extension endeavored to make homemakers into self-assured consumers by encouraging them to defer to the opinions of university-trained experts. Mrs. Ellsworth Richardson stated that modern society had become far too complex for women to manage their homes without the expert advice of the Extension Service and women's organizations. In 1938 Mrs. Eugene Cutler, president of the Iowa Federation of Womens' Clubs, declared that home economists and the Extension Service had transformed homemaking and childrearing practices across the state in only a couple of decades and observed that women no longer gathered together in sewing circles to exchange tips about childrearing, clothing, or nutrition with one another but now deferred to the expertise of home economists.[62]

Really Rural Plays

The Country Life movement sought not only to make life better for farm women but to alleviate some of the cultural privation that farm families experienced. Efforts to alleviate the tedium of rural life and promote home economics education converged in the Extension Service's Little Country Theater project, created in 1920 to uplift small-town and rural society by encouraging residents to stage entertaining and instructive plays in their own communities. As the *Homestead* observed, the Little Country Theater demonstrated "how any rural community can put on its own amusements," and so was "of interest to every farming community in the state where the question of entertainment has been a puzzling one." The Little Country Theater movement, however, soon found itself ensnared in a debate over the relative importance of supplying entertainment and education to rural Iowans, a debate between those who sought to produce dramatic plays and those who considered the theater another vehicle for disseminating information about scientific household management.[63]

Iowa State College's Extension Service introduced the Little Country Theater program to communities across the state, supplying county extension agents with plays deemed suitable for local theatrical troupes and replying to dozens of requests from rural women for scripts and staging suggestions. The state fair offered an ideal venue to publicize the Little Country Theater movement, and a troupe of Iowa State students performed at the fair in 1921.[64] Initially, Little Country Theater companies performed free of charge before the fair's enormous grandstand. In 1923 they were moved to the Women's and Children's Building so that the amphitheater could host the growing number of more popular and profitable concerts, auto races, and disaster spectacles. Unfortunately, the din from the fair's Midway often drowned out the muses: in 1926 the college's theatrical company announced that henceforth it would perform only comedies, because the boisterous crowds and sideshow barkers outside the auditorium detracted from tragedies and other serious dramas.[65]

Audiences at the Little Country Theater performances enjoyed a variety of plays, ranging from educational skits, in which characters such as Minnie Minerals and Polly Protein earnestly instructed the audience about the benefits of niacin, to Chekhov. Many extension agents viewed the Country Theater project primarily as a means to promote scientific household management, rather than a form of entertainment or a means to foster culture in rural communities. Plays written by the Extension Service addressed issues of homemaking and health, such as *Good Taste in Dress*, *The Becoming Hat*, *What to Eat*, *Home Management*, and *The Joy of Walking*. These plays typically shared a simple, familiar plot: a beleaguered housewife is rescued from drudgery and her disheveled home miraculously transformed when she adopts the homemaking advice offered by a perky extension agent.[66]

What Every Woman Knows, a frequently produced Extension Service play, typifies the genre and also suggests the skepticism with which extension agents were greeted in many rural homes. As the play opens, Len, a farmer, scoffs at a newspaper advertisement for a lecture on household management by the county extension agent, grumbling that this "old maid" will only "put a lot of new-fangled notions in the heads of all you women."[67] His wife, Molly, defiantly insists on attending the lecture, leaving her husband in charge of the home and their child. As she departs, he blithely assures her that "a man who has run a thousand-acre farm for ten years can manage a house and kid" for a few hours. Fittingly, Len receives his comeuppance. During Molly's brief absence the home is turned topsyturvy, and she returns to find Len at his wits' end and a newfound convert to the science of home economics. "Molly," he says contritely, "I want you to be sure and go to these meetings about home conveniences, and I'm going to town tomorrow and buy a new washboard, and a broom, and a mop, and. . . . " The audience applauded Len's newfound appreciation of the challenges of being a homemaker and readily accepted the importance of improving conditions in the farm home. *Dollars and Sense*, another Extension Service play, emphasized that the household was indispensable not only to the family's well-being but to the farm's economic viability and urged farmers not to neglect the home when allocating the family's income. Disagreements over expenditures and priorities were commonplace in many rural households, as men often insisted on spending money on farm implements and tools, while neglecting the supposedly "nonproductive" household. Efforts to persuade farmers to devote more of the family's income to the home became a staple of home economics skits.[68]

The state's agricultural periodicals generally gave rave reviews to the didactic home economics skits and urged Little Country Theaters across the state to produce plays that would educate farm families rather than give rural thespians an opportunity to display their talent. Reviewing the Little Country Theater's program at the 1923 fair, *Wallaces' Farmer* lamented that too many of the productions were not truly "rural plays." Rural plays, the magazine wrote, might not be particularly sophisticated or entertaining, but they taught explicit moral and practical lessons. The magazine applauded the fair's home economics skits, *Queen of Foods*, *Dollars and Sense*, and *Home Harmonies*, as examples of "really rural plays" and suggested that plays should have large casts, so as "to use as much as possible of the talent available rather than to train a few people in acting." Plays should be instructive and inclusive, rather than seek to cultivate actresses and actors.[69]

The organizers of the Little Country Theater project, however, were less interested in using the stage to promote home economics. They encouraged rural and small-town Americans to produce and attend their own theatrical entertainments, providing their neighbors an inexpensive, homegrown alternative to movie theaters and creating a vibrant local culture. Many participants in Little Country Theaters around the state apparently felt likewise: most of the local theatrical groups who requested scripts and staging suggestions from the Extension Service expressed a strong preference for dramatic plays over one-act home economics lessons. Little Theater project director Frederica Shattuck of Iowa State College worried that being relegated to the Women's and Children's Building would impede her effort to promote the creation of local theatrical companies, because "practically the entire audience is made up of women."[70] Shattuck hoped to provide rural Iowans with an inexpensive, accessible diversion, to acquaint them with dramatic literature, and to enable small town and rural residents to experience the thrill of performing onstage. Shattuck discouraged one local theater company from producing *Dollars and Sense*, confiding that she "had no objection to the lesson, of course, but often plays written for the purpose of teaching a lesson are not really well written and not very entertaining. They are likely to be dull."[71]

The Little Country Theater movement scored a few hits. Throughout the 1920s and 1930s, county extension groups and Farm Bureau chapters throughout the state formed Little Theater companies and staged their own performances. In the 1930s, more than eighty local theatrical groups presented several hundred plays annually. Proponents of the Little Country Theater movement were keenly aware that the lack of entertainment in rural areas ranked among the leading sources of dissatisfaction with rural life. Yet their hopeful belief that the remedy for rural isolation lay in encouraging rural Americans to provide those diversions themselves, however appealing, scarcely posed a serious challenge to the booming entertainment industry. Both at the fair and in theaters across the state, commercial entertainments, not locally produced plays, took the lead role in providing diversion for farm families. While Little Country Theater troupes performed before small but appreciative audiences in the Women's and Children's Building, the turnstiles at the fair's amphitheater were spinning, as thousands streamed in to enjoy the races and grandstand shows. Gargantuan spectacles, variety shows, and other amusements before the fair's grandstand, not didactic plays staged by civic-minded thespians, provided the fair's main attraction in the 1920s and 1930s. *Billboard*, the entertainment weekly, declared confidently that the growing number of Little Country Theaters posed no threat to the show business, but ironically helped to inculcate "urbane tastes and understanding" among rural Americans, making them even hungrier for high-class, professional entertainments. The Little Country Theater, according to *Billboard*, was not the professional showman's rival but his advance agent.[72]

By the 1930s, radio and movies had already transformed popular entertainments in rural America. *Variety*'s legendary headline, "STICKS NIX HICK PIX," encapsulated rural Americans' understandable desire to enjoy the same entertainments available to urbanites. No longer content to sit through second-rate or second-run movies at their local theaters, residents of farms and small towns were now expected to see the same movies that screened in big cities.

In 1938 Mrs. Eugene Cutler, of the Iowa Federation of Women's Clubs, observed that the tremendous influence of mass entertainment and mass communications was evident even in the state fair's 4-H demonstrations and educational exhibits: "The radio, motion picture, the newspaper, the magazine, the automobile and countless other things . . . produce a restless type of audience—an audience that will remain quiet only as long as the program from the platform is interesting, new and colorful." In order to hold patrons' interest, the fair "must entertain as it educates." Now that even rural Iowans were accustomed to enjoying professional entertainment, the fair's educational features had to be diverting in order to capture and hold the audience's attention. Cutler's words attest to a significant transformation in rural life in the 1920s and 1930s: as automobiles, radios, and motion pictures reduced the isolation of rural life, even farm families became more enmeshed in American consumerism and mass culture.[73]

Commercial entertainments and home economics extension both rapidly diminished the gulf between rural and urban Americans in the early twentieth century, and showmen, even more than home economists, transformed the character of state and county fairs. The boundary between the fairs' entertainment and educational offerings had never been distinct, and the fair had always offered a mixture of entertainment and education. But this boundary became even blurrier in the early twentieth century as entertainments began to occupy a much more prominent place on the fairgrounds. Country Lifers, home economists, and extension agents transformed the fair's educational exhibits in an effort to make farm life more attractive. By teaching farm families to become astute consumers, attempting to satisfy their desire for entertainment, and seeking to bridge the gap between rural life and town life, they strove to preserve country life by acquainting rural Iowans with the norms and styles of a rapidly urbanizing society. Farmers, agricultural journalists, home economists, and fair men would all continue to proclaim the fair's fidelity to agriculture, but rural life and the fair, like the nation, had been irreversibly transformed by the seismic shift toward urbanization, industrialization, and consumer culture.

1 The fairgrounds, looking up Grand Avenue toward the Exposition Building,
now known as Pioneer Hall.

Iowa State Fair Collection, SHSI, Des Moines (loose photos).

2 Sideshows, inside the fairgrounds.

Iowa State Fair photographs, Charles Buswell Collection, SHSI, Des Moines, box 6.

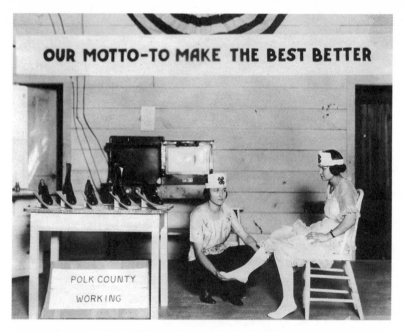

3 A 4-H demonstration exhibit: Ruth gives Pearl some foot exercises.

Josephine Arnquist Bakke papers, Special Collections, Iowa State University Library.

IN THE STOCK PAVILION

4   Inside the Livestock Pavilion, 1906.
    Iowa State Fair Collection, SHSI, Des Moines, box 4, Cattle folder.

THE WORLD'S GREATEST OPEN AIR EXHIBITION

TOO BIG FOR TENT OR ROOF, HEAVEN'S LOFTY DOME ITS ONLY CANOPY

EVERY EVENING
In front of the New Steel Amphitheater

Depicting how the modern air ship will be used to raze cities
and destroy whole armies in future wars

Each performance terminates with a beautiful display of Pain's Fireworks

....NEW FEATURES LARGER AND BETTER THAN EVER....

*This performance will be preceded by a Grand Concert by Liberati's
Concert Band—45 people, 4 voices*

DOUBLE PROGRAM FOR ONE ADMISSION

Admission 25c, including Reserved seat 50c, Box seat 75c
Reserved seats may be purchased for any performance after August 21st at the following places;
Oleson's Drug Store, near street car waiting room
West & Kitchen Drug Store, East 5th and Locust Sts.
Treasurer's Office, Administration Bldg., Fair Grounds

5   The future of warfare: "Battle in the Clouds."
Iowa State Fair Official Program, 1909: 18.

6 "An entire city blown up": "The Fall of Troy."

Iowa State Fair Premium List, 1927: 237.

Scene from "A Night in Bagdad"

Many strange and weird sights will greet fair visitors this year in the unusual fireworks extravaganza, "A Night in Bagdad," which will be produced in front of the grand stand every evening of the exposition.

Above is a representation of one of the episodes in the spectacle which recalls vividly the ancient tales of the Arabian Nights. The settings are said to be unusually brilliant and the action of the spectacle especially original and thrilling.

7 A clash between good and evil genies: "A Night in Bagdad."

Greater Iowa, May 1928: 5.

8   The fair's entertainment zone: the Midway at night in the late 1930s.
Iowa State Fair Collection, SHSI, Des Moines, box 5, folder 1.

9   The pinnacle of the fair's exhibits: fine arts.
Iowa State Fair photographs, SHSI, Des Moines, box 1, Exhibits folder.

10   When tillage begins . . .

11   . . . other arts follow. Dan Rhodes and Howard Johnson's mural.
     Greater Iowa, July 1939: 4–5.

12   A century of history: The fair's gate in 1938.
     Greater Iowa, July 1938: 3.

A Bumper Crop of Entertainment

FAIR MEN, MIDWAYS, AND SPECTACLES

"We are very gravely informed," *Billboard* announced with mock solemnity in 1915, "that the original mission of our state fairs has been largely accomplished, that we have now reached and are entering upon a new epoch in which a new kind of educational service is demanded of these institutions and that this demand will grow steadily more and more insistent." The editors of "the show world encyclopedia" were unimpressed by learned professors of home economics and animal husbandry. "There is a place for the agricultural college," the magazine declared, "but it is not on the fairgrounds." Eliminating amusements from fairs would not revive the exhibitions' popularity: "It is the surest way to kill fairs."[1]

For decades, the state fair's secretary had performed a balancing act, seeking to admit enough amusements to make the fair alluring and profitable without causing controversy over tawdry or crooked sideshows. The fair could survive a little grumbling about the impropriety of burlesque acts or gambling from agricultural journalists or clergymen as long as it did not provoke a public outcry or the wrath of a state legislator. But as the fair began to book more games, shows, and rides to attract a crowd, its officers recognized that they were now in competition with other commercial amusements for the public's attention and money. As fairs became an important part of the show business, fair secretaries found themselves competing with vaudeville, movies, and radio, and they worried that agricultural fairs now seemed staid and antiquated compared to other amusements.

Although amusements started to receive top billing at the fair in the early twentieth century, the fair's managers did not suddenly throw open the gates to any showman eager to pitch his tent on the grounds. As entertainments became a bigger part of the fair's bill, they simultaneously became big business, and fair men, booking agencies, and

large show business companies cooperated to regularize and clean up
the fair and the show business. As the show business became big business, larger, more successful showmen and companies sought to drive
some of their smaller, less reputable rivals out of the industry in order
to improve the seedy image of carnivals and sideshows. As Jackson
Lears has observed, many middle-class and elite Americans strove to
"contain" the unsettling carnivalesque elements in their culture in the
late nineteenth and early twentieth centuries in order to consolidate
an economy and society increasingly dominated by corporations and
committed to bureaucracy, rationality, and efficiency.[2]

The growth of show business did not end the debate over the relative place of agriculture and entertainment at the fair. Well into the
twentieth century, critics of entertainments continued to depict showmen as villains who corrupted the fair and swindled its patrons, and
maintained that education constituted the fair's only legitimate function. Critics of amusements charged that lurid sideshows and crooked games undermined the fair's educational mission, viewing them as
harbingers of a culture devoted to leisure and self-indulgence rather
than labor and productivity. As always, debates over the fair's agricultural exhibits and entertainments spoke to larger concerns about the
future of the Iowa's economy and culture.[3]

Showmen and Fair Men

In the late nineteenth and early twentieth centuries, the fair business
changed rapidly, as fairs became important venues for the outdoor entertainment industry and "fair men," not agriculturists, began to manage fairs. The 1893 Columbian Exposition marked a watershed in the
history of fairs and the show business, but the emergence of the show
business as big business predated the World's Fair's legendary Midway.
The outdoor entertainment industry began to emerge as a big business
in the 1880s, and the management of fairs immediately became an important branch of this new industry. Like other businessmen, fair men
and showmen availed themselves of the advantages of consolidation
and economies of scale, forming larger companies, creating trade associations, and linking separate fairs into circuits. In 1884 fair men met
in St. Louis to found their own trade organization, the International
Association of Fairs and Expositions. The following year, fair men and
showmen created a separate midwestern fair circuit, the Western Fair
Managers Association, which sought to regularize the process by which

fairs booked their entertainments and spare fairs from competing with one another for dates and bookings. The association merged with its eastern counterpart in 1896 to form the American Association of Fairs and Expositions.

In 1913 entertainers founded the Showmen's League of America, electing Buffalo Bill Cody as its first president, and in 1918 the National Outdoor Showmen's Association. These organizations of fair men and showmen, which enabled fairs to avoid scheduling conflicts with one another, convened annually in Chicago so fair men could meet to exchange information and book acts for the next summer's exhibition. Midwestern state and county fair managers also banded together to create fair circuits, scheduling their runs consecutively and booking entertainment companies to play each fair in the circuit. Assured of steady work and comparatively short distances to travel between engagements, entertainers in turn agreed to perform for reduced fees. Forming fair circuits to book attractions helped transform fairs from individual exhibitions into venues in the outdoor amusement industry. Formerly, fairs had been exhibitions run by agriculturists and local boosters; now fair men and showmen banded together to create the fair business.[4]

Consolidation within the show business sharply reduced the number of venues for solo acts. Independent showmen continued to play the fair for years, but the days were numbered in which the fair secretary's mailbox was filled with stacks of correspondence from unknown impresarios seeking to perform at the fair. Forced to sift through dozens of advertisements for unfamiliar acts ranging from acrobats and balloonists to trained ostriches and trotting moose, the fair's secretary frequently complained that booking attractions had become a time-consuming nuisance. At the outset of the twentieth century, fair secretaries began to contract almost exclusively with large amusement companies and booking agencies, which offered a variety of dependable, crowd-pleasing acts for the fair. Independent showmen, recognizing that their opportunities to secure bookings were rapidly diminishing, had strong incentive to join these larger amusement companies. As the outdoor amusement industry grew and fair men and showmen consolidated their enterprises, the business of entertainment became much less unruly. The fair business was now part of the show business, and show business had become big business.[5]

Initially, fairs had been overseen by agriculturists; now the fairs' managers proudly called themselves "fair men," and the fair business became a distinct branch of the show business. The successful fair man still had to know the difference between a Hampshire and a Duroc boar, but he also had to possess a showman's keen instinct for attracting a crowd and putting it in a free-spending mood. Fair men could quote Thomas Jefferson about the dignity of farming and discuss crop prices, but at heart fair men were showmen, and they believed that entertainments attracted a crowd to the fair. *Billboard* spoke for fair men in 1898, asserting that any attempt to convert the fair "into an educational institution or a straight-laced department of the church . . . will fail miserably. It is not and never was intended to be either a sermon or bald lesson. The shows, music, races and legitimate games are as much a part of the fair as the exhibits themselves."[6] The following year, *Billboard* advised fair men on how to run successful modern fairs that would entice city dwellers and country folk alike. The most important lesson for any fair manager was simple: "The day of the purely agricultural fair is past."[7]

Spawn of the Chicago Midway

The combined wallop of the depression of 1893 and competition from the Columbian Exposition left the state fair deep in debt in the 1890s. For years, virtually every discussion of entertainments at state and county fairs would be tinged by memories of the exposition. The state fair's new secretary, P. L. Fowler, foresaw an "uphill business for all fairs . . . until the World's Fair has been forgotten." The Columbian Exposition, he observed, had "so over-shadowed the Fair business that we will have to look to some other line, than what we have had here-to-fore." Fair secretaries across the nation conceded that the World's Fair had set an unmatchable standard for exhibitions and outdoor amusements but felt impelled to emulate the exposition to whatever extent they could. Dazzled by the exposition's celebrated Midway, state and county fair secretaries across the nation created their own "Midways" in imitation of the entertainment zone at the World's Fair and painted their wood-frame fair buildings white to emulate Chicago's famed White City. After the World's Fair closed, state and county fairs, especially in the Midwest, found themselves inundated with advertisements for acts from the exposition's Midway and the booming outdoor entertainment industry, and the state fair's secretary reported a substantial increase in the number of shows, reputable and otherwise, seeking space on the grounds. Fair men began to admit more showmen into the grounds and to feature them in the fair's advertising in an effort to make fairs more alluring.[8]

In the spring of 1894, secretary P. L. Fowler cast about for ideas for pulling the state fair out of debt. Fowler considered himself a fair man, not an agriculturist, and believed that entertainments held the key to the fair's success. He considered booking the Pain Pyrotechnic Company to stage a fireworks spectacle at the 1894 fair in the hope of duplicating the success of "Pompeii" the previous year, but booking such an expensive entertainment proved too great a risk for the society, which could ill afford to slide deeper into debt. Fowler opted instead to book an array of small acts, hoping to create "a continuous attraction during the whole Fair." The fair's attractions included two balloon ascensions—one in which human daredevils and dogs parachuted to the fairgrounds, the other featuring a man shot from a cannon suspended from twin balloons (!)—a trained dog act, sharpshooter (and Buffalo Bill's niece) Lillian Cody, an educated horse, and Civil War veteran H. B. Hendershot, who billed himself as "the original Drummer Boy of the Rappahannock." Although the 1894 fair did not have to compete with the Columbian Exposition, the depression continued to deflate farmers' spirits and wallets, and when the fair's gates closed, the society remained mired in debt.[9]

While booking attractions for the 1895 fair, Fowler observed matter-of-factly that "we are in the show business," and so must be responsive to the tastes of the fair's patrons.[10] As the fair's opening day neared, Fowler informed the state's newspapermen that the upcoming exhibition would offer more amusements than its predecessors:

ATTRACTIONS FOR THE FAIR

The State Fair Goes Into the Show Business for All It Is Worth.

The state fair is making an effort this year to get into the show business in earnest, since it has to compete with all manner of shows for public favor, and it has been demonstrated that people will not support a purely agricultural exhibition of the magnitude of the Iowa State Fair. The side attractions at the fair this year are several times greater in number and immeasurably greater in drawing strength than at any previous fair.[11]

The fair's bill of entertainers sounded downright intellectual, including Professor Herbert La She, a high wire aerialist; Professor K. P. Speedy, a daredevil who plummeted seventy feet into a tank of water only three feet deep; and Professor Hunt, with his troupe of twenty-one trained canines. For lowbrows the fair offered less cerebral-sounding acts, including the acrobatic Rozella Brothers, balloon ascensions, bicycle races, bagpipers, trapshooting, and the usual brass band.[12]

In 1896, three years after the Columbian Exposition, Secretary Fowler had the fair's buildings painted white and billed as Iowa's own White City, but no amount of whitewash and advertising could transform the state fair's modest wood-frame exhibition buildings into the exposition's neoclassical cityscape. When the fair lost money for the fourth consecutive year, even the agricultural society's directors abandoned their usual boosterism and wondered whether the fair could be made financially viable. After struggling to keep the state fair solvent amid a wrenching depression, widespread political and economic discontent among farmers, and endless disputes over the fair's proper role, many fair men fretted that something was fundamentally wrong with the fair and pondered what, if anything, could be done to fix it. At the agricultural society's annual meeting in 1897, John Cownie of South Amana delivered an address in which he asked, "What are the causes of the lack of interest in the state fair?" Fairs, he suggested, confronted several challenges. The fair's agricultural exhibits had waned in importance, while its entertainments had become more prominent. Farmers no longer needed to attend the fair to learn about livestock, new machinery, or scientific agriculture, because breeders, dealers, and farmers' periodicals now reached them in their own locales. The stupendous World's Columbian Exposition had rendered the state fair's exhibits forever "meager" by comparison. Fairs, he concluded, must either change with the times and find ways to attract patrons, or cease to exist.[13]

The economic turmoil of the 1890s reminded agriculturists and fair men that the state fair remained extremely vulnerable to depressions, competition from rival attractions, and bad weather. The agricultural society simply could not host a successful fair, maintain and improve the fairgrounds, and pay its expenses solely from its gate receipts. Even one year in the red left the society without money to book and advertise the following year's exhibition. Additionally, as Secretary Fowler observed, fairgoers now demanded "a much higher order of exhibit" at the state fair.[14] He conceded that agricultural society's responsibilities were too great to be shouldered by a private organization, especially one utterly dependent on the fair's gate receipts for revenue. The society's president declared in 1898 that the annual fair and the agricultural society were no longer adequate to educate farmers and foster economic development. As he put it, Iowans "demand more than an annual fair."[15]

The 1899 fair attested to the fair's waning importance as an educator and as an exhibition of innovative agricultural techniques. After casting about for an inexpensive way to promote the fair, Fowler billed it as the "Closing Century Exposition," which would feature historical displays of obsolete implements, overfattened livestock, and other relics from the pioneer era, as well as exhibits designed to "show the development to the present time." The fair's advertisements hailed the exposition as "intrinsically better worth seeing than its predecessors," noting that it furnished "an interesting contrast between the appliances, processes, and methods of life today and those of the earlier years of the century."[16]

The Closing Century Exposition was in part a ploy to lend the fair some allure on the cheap, without booking expensive entertainments, but it also signified a telling shift in the fair's role. In earlier decades the fair was billed as a showcase for displaying and promoting innovations in agriculture and machinery, and the fair's managers discouraged the exhibition of obsolete practices and items. As the pioneer era receded into the past, Iowans became wistful about the "old settlers" who had staked the first claims, broken the prairie, built homes, and founded communities. The fair's Closing Century Exposition in 1899 was Janus-faced, gazing wistfully toward the past and looking hopefully toward the future. New implements and farming techniques were still on display at the 1899 fair, but the fair had also become nostalgic. The fair's newfound orientation toward the past resulted from Iowans' growing awareness of the state's history, which was now long enough for citizens to take stock of its progress since the pioneer era. But the fair's backward-looking gaze also attested to its diminishing importance as a site for agricultural innovation. After less than a half-century, the fair had become a repository of tradition, rather than an agent of progress.

For decades the society's officers had insisted that their organization deserved to be made a full-fledged state institution and to be assigned a larger role in promoting Iowa's economic development. After the Closing Century Exhibition, the officers of the Iowa State Agricultural Society requested that the state legislature disband the organization and replace it with a full-fledged department of agriculture to oversee the state's principal industry. The new department would continue to hold a state fair, but, as P. L. Fowler noted, "that would be among the least of its duties." In March 1900 the General Assembly discontinued the Iowa State Agricultural Society and established the Iowa Department of Agriculture, which took over the society's functions. Managing the fair was only one of the department's duties, and it would occupy only one chapter in its hefty *Iowa Yearbook of Agriculture*, a compendium of statistics and reports on the state's crops and livestock.[17]

When the state government assumed responsibility for the fair in 1900, opponents of entertainments spied an opportunity to lobby for an exclusively agricultural fair and eliminate amusements from the annual exhibition. They contended that the state government should not be engaged in the show business or risk scandal by permitting disreputable amusements at the fair. "The fair has reached the point where it can live independent of this sort of revenue," declared the *Homestead*, criticizing the amusements at the 1902 fair. The Fair Board's president, J. C. Frasier, echoed this sentiment, announcing that "the time has arrived when the Iowa State Fair should exclude all side shows from the grounds. The only excuse that ever could be made for them was the need of the revenue received, but that is no longer an excuse." The State Board of Agriculture, which now oversaw the fair, rejected Frasier's call to exclude sideshows, resolving "that no show of an objectionable nature" would be permitted at the fair but contending that "there may be shows that are entirely unobjectionable, that are not only entertaining, but instructive, and add to the attraction of the fair." The board advised the fair's secretary to "use due discrimination in the selection of these shows." A journalist covering the 1903 fair reported that the ragged tents of traveling showmen had been supplanted by "clean-looking booths" and noted "a gratifying absence, almost completely, of the usual Midway Plaisance features."[18]

Still, agricultural periodicals and other critics of amusements contended that the fair now had the sanction and support of the state and should exclude entertainments altogether. The *Homestead* reported that "all forms of cheap side shows or anything bordering on the vulgar-sensational will be prohibited by the authorities at all future fairs." When the 1904 fair turned a profit, *Wallaces' Farmer* attributed its success to the fair secretary's decision not to book "the young lady who used to stand on a platform ornamented with snakes of various sizes and colors" and "the Senegambian who poked his head through a hole in a tent and allowed boys to win a prize for hitting him with rotten eggs," and to offer instead only "clean" entertainments.[19] After a decade of being "submerged in the flood of filthy side shows, spawn of the

Chicago Midway," the magazine declared, fair men had learned that they could run a successful exhibition without allowing showmen to corrupt and swindle the fair's patrons.[20] But the fair had scarcely been devoid of entertainment. The Fair Board may have barred "objectionable" shows from the fair, but Secretary J. C. Simpson noted that he had spent $5,000 more on entertainments in 1904 than he had the previous year. "To this extra effort in providing star attractions," he stated, "we attribute the success of the fair. . . . All education and no amusement makes the fair a dull place."[21]

The Fair Business

At the outset of the twentieth century, fair men still booked entertainments by contracting with individual showmen, who filled the secretary's mailbox with gaudy flyers, emblazoned with superlatives touting their freak shows, snake acts, high-wire routines, trained animals, and myriad other attractions. Both fair secretaries and managers of large entertainment companies had considerable incentive to rid their business of the scads of small, independent sideshows. Booking acts individually required countless hours of labor by the fair's secretary, who sifted through bushels of correspondence from unknown showmen. Some of these itinerant showmen proved notoriously unreliable, failed to live up to advance billing, or even failed to appear if they received a more lucrative booking elsewhere. As the secretary lamented, booking small acts was a gamble, as he had no guarantee that showmen would adhere to their contracts, "so we have simply to take our chances with them."[22]

Because contracting with individual showmen was a dicey proposition, fair men increasingly turned to booking agents and large carnival companies to schedule entertainments for their fairs. Beginning in 1901 the Iowa State Fair arranged some of its entertainments through booking agents; within a few years, it relied on them for nearly all of the fair's attractions. Booking agents promised to supply the fair with a variety of "clean" acts designed to appeal to virtually every taste and guaranteed to perform as scheduled.[23] Attractions at the 1908 fair included:

- a trained animal circus
- the ossified Kilpartick
- Alice, the wonderfully hairy little woman
- Randiou, the Hindoo who, without feet or hands, rolls cigarettes

and ten pins with equal facility, and who does everything from
magic to music
- the six Bachman glass blowers
- the Great London Ghost Show, which is made up of astonishing
illusions
- a penny arcade or vaudeville
- an oriental show, which will be clean and devoid of the real dance
[belly dancing]
- Serpentina, who is a freak
- and other features.[24]

In 1909 *Billboard* proclaimed that consolidation within the show
business had transformed fairs from ragtag exhibitions into reputable,
well-managed venues for the show business, and that fair secretaries
were now hired "as much because of their knowledge of the amusement
business as for their ability to handle the agricultural fair." Until recently, the magazine wrote, fair men had booked sideshows "without much
thought as to system, merit or their real quality." Astute fair men now
booked only "the best to be obtained in the way of attractions, which
have come to be dignified by the name 'concessions,' the side show having been relegated to the past." Cooperation enabled fair managers "to
avoid the pitfalls into which they formerly fell as individuals, until now
fairs are conducted in a business-like manner that completely obliter-
ates the element of chance." As an agent for one of the nation's largest
booking agencies wrote in 1910, choosing the fair's acts was no longer
a crapshoot but had become as simple and dependable as ordering a
sofa from Sears, Roebuck: the modern fair secretary simply "orders
the numbers on the program as he does the furniture in his house—by
catalogue."[25]

The effort to systematize the fair business did not occur simply because it spared fair managers headache and expense but in the interests of the outdoor amusement industry. Fair men and showmen alike
blamed small, sleazy acts for sullying the reputation of their business,
diminishing attendance at outdoor shows, and exposing the entire outdoor entertainment industry to potential backlash from state and local governments. W. D. Ament, proprietor of the "Great London Ghost
Show," complained to fair secretary John Simpson in 1905 that disreputable attractions "have ruined the business for clean Legitimate Shows
on nearly all the Fair Grounds from St. Paul to the Gulf of Mexico."[26]
In 1910 *Billboard* urged honest showmen to combine in order to drive

their corrupt peers out of business, for "unless is exterminated the unscrupulous individual who ekes out his existence by preying upon everybody with whom he comes in contact, using methods that are below-board to carry on his illegitimate enterprise, the fond hopes of many for the future of the business will be torn asunder." "Legitimate" showmen combined to force tawdry showmen to clean up their acts and join larger carnival companies or drive them out of business altogether.[27] Fair men insisted that fairs should be managed like any large business, and they waged a public relations campaign to persuade Americans that fairs were no longer seedy carnivals but large, well-run, respectable businesses, no different from any other enterprise.[28]

The components of the emerging outdoor amusement industry— booking agents, large carnival companies, fair circuits, and associations of fair managers—combined to restrict the number of venues for smaller showmen, especially those whose acts could be accused of being risqué or fraudulent. Profitability and predictability, as much as prudery and decency, however, prompted fair men to clean up outdoor amusements. In 1909 *Billboard* boasted that sideshows had "undergone a complete metamorphosis during the past few years." The "rag front" attraction and "fake show . . . engaged haphazard and without care or thought, either as to its moral or edifying influence," had been replaced by the gleaming, modern show "with its performance of real merit." Freaks, burlesque shows, and rigged games "have been relegated to the annals of amusement history, if they are to be remembered at all."[29] By driving small showmen out of the fair, large show business companies and fair men sought to insulate themselves from charges that sideshows corrupted farm youth and hastened the exodus from the countryside to the city.[30]

Critics of entertainments sometimes insisted that games and shows be cordoned off from the rest of the fair and surrounded by a high board fence, so they would not contaminate the exhibition. On the other hand, enclosing sideshows within a fence or billing them as a "midway" often caused their opponents' imaginations to run riot. Throughout the fair's history, agriculturists and fair men had attempted to draw a boundary, both physically and symbolically, between the fair's legitimate exhibits and its entertainments. Many sideshows had literally been kept outside the grounds, while others were permitted within the grounds but confined to an area separate from the agricultural displays. After the 1893 Columbian Exposition, the fair's Midway became the site of most of its games, shows, and rides. But showmen sometimes defied the fair's attempt to confine them to a separate area and frequently employed pitchmen to roam the fairgrounds touting the Midway's attractions. Showmen at the 1900 fair refused to remain within their allotted space on the west side of the grounds, staging a daily "Midway parade" around the fairgrounds, led by a barker proclaiming the Midway's wonders, who was trailed by "several dark-hued men tricked out in Oriental costumes, possibly an Indian or two, a few dancing girls, one or two weird musical instruments and a trick bear of the cinnamon species." The parade captured the attention of hundreds of spectators, many of whom, like the proverbial rats of Hamelin, could not resist trailing the parade as it wound its way back into the enclosure.[31]

When booking amusements for the 1908 fair, the secretary stipulated that "they are not to be placed in an enclosure, nor are they to be advertised as a Midway," because doing so would only antagonize critics. On the other hand, the *Homestead* argued, if the fair must have amusements, "then have them segregated or districted, where they will not disturb those who come to enjoy the fair and not to see a side show." When the fair enlisted landscape architect O. C. Simonds to design a comprehensive plan for the fairgrounds in 1910, Simonds designated the area directly east of the fair's grandstand and racetrack for amusements; inevitably, this area was dubbed the Midway, home to a cacophony of barkers and calliopes, and an equally discordant debate over amusements.[32]

Wallaces' Farmer called the fair's designation of an official space for amusements "a decidedly backward step" and complained that the Midway included several shows "that had no business on the Iowa State Fair grounds, or on any fair grounds frequented by decent people." Confining shows to a specific portion of the grounds did not prevent them from contaminating the entire fair, according to the magazine, which declared that "the midway should be done away with, and the stream of filth which flows from it should be dammed up once and for all." The following year, the *Homestead* charged that the fair had sunk even lower by booking "Streets of Cairo," named for the legendary belly dancing act at the Chicago World's Fair, which "flaunted its indecency in the face of the public throughout the entire week." Before each performance, the dancers appeared onstage in front of the tent, "and by muscular gyrations and surreptitious winks allured men inside."[33]

The upstanding journalists, clergymen, and legislators who sought to make the state fair a shrine to agricultural bounty often viewed Midway acts, especially dancing girls, freak shows, and games of chance, as the embodiments of an insidious threat to the state's economy and culture. They charged that shows and games undermined values of hard work and frugality and ran counter to the fair's effort to promote rural life and accused the fair of tolerating amusements in a misguided attempt to attract city dwellers, as though entertainments held no allure for rural Iowans. *The Des Moines Register and Leader* lamented in 1911 that the state fair's bill of entertainments grew longer each year because shows, games, and spectacles offered "the only way to insure the attendance of a great majority of the city folk," who had no interest in viewing exhibits of livestock and implements.[34] A few years later the *Homestead* similarly claimed that "fully nine-tenths of the people enjoying the horse racing and the vaudeville acts were town folks," while the farmers were "engaged in increasing their knowledge of farming and rural conditions generally."[35]

Billboard, "the showmen's bible," scoffed at high-falutin' rhetoric about fairs' agricultural and educational mission. Fairs were part of the amusement industry and would succeed or fail according to the quality of their attractions. Clean, well-run amusements, according to *Billboard*, were indispensable to the fair, whatever outdated ideas a few "fossilized" critics might believe.[36] Showmen and fair men denied that shows and midways corrupted fairs or impeded their educational mission. While fairs might convey a few useful lessons to their patrons, *Billboard* insisted in 1915 that "the most important mission of the fair remains today exactly what it has been from time immemorial, i.e., to make a holiday season for the people—a fall festival—to which the promise of entertainment lures, the prospect of reunion invites, the outcome of competitive exhibits draws and the opportunity of gaining information and acquiring new knowledge is a quite secondary, although important matter." According to the magazine, "A modern state fair is a celebration, a festival, a vacation, a recreation. Its essential spirit is entirely foreign to the morbidly serious purpose which our toiling forefathers had in view when they brought fairs into existence."[37] Fair men believed the "showmen's bible": entertainments were the fair's main draw. As one fair man bluntly put it, "racing and vaudeville and amusements of all sorts are the features that attract the crowds to the county fair. People do not come to the county fair for education; they come for amusement. This education talk is all a joke."[38]

By the 1910s the state fair spent tens of thousands of dollars annually to book its Midway, grandstand acts, and sideshows, and the fair's advertisements emphasized its attractions, not its livestock competitions. In 1919, for example, the fair contracted with C. A. Wortham Shows, which supplied a veritable cornucopia of entertainment, including "The Big Wild West Show and Indian Congress," "Don Carlo's Dog, Pony, and Monkey Hotel," "Joyland Follies" (a troupe of dancing girls), "Captain LaDare," "The Girl from Delhi," auto daredevils, "Stella and Her Submarine Girls" (an aquatic act), "Genuine Filipino Midgets," "The Popular Athletic Hippodrome," clowns, tumblers, a wax museum, several carnival rides, "Danger" (a glimpse of a Chinatown opium den and "the dangerous results of the use of deadly drugs and quack medicines"), and "Pajana's Mammoth Enlarged Circus Side Show" (a collection of human and animal curiosities). Livestock contests and 4-H exhibits still attracted farmers young and old, male and female, but the fair's new image as an annual carnival and vacation owed more to P. T. Barnum than to Thomas Jefferson.[39]

Crooked Carnivals

Longstanding tensions between rural and urban America became unusually bitter amid the economic and cultural upheaval of the 1920s. Prices for agricultural commodities plummeted after World War I while the industrial economy generally soared. Grain markets tumbled a decade before the stock market crash of 1929, and many farmers were already deeply mired in depression when the economy crashed down atop them at the end of the decade. The ravaged farm economy only added to the widespread fear that rural life no longer held much appeal for farmers or garnered much respect from Americans. The growing availability of automobiles, radios, motion pictures, and mass-circulation magazines disseminated cosmopolitan culture and values to small towns and rural areas. Paradoxically, these new technologies and mass media simultaneously reduced the differences *and* fueled tensions between rural and urban Americans in the 1920s. The contrast between rural and urban life was especially evident during fair week, when farmers and city dwellers alike congregated on the fairgrounds in Des Moines.

County fairs, generally considered an index of the well-being of the countryside, dwindled in the 1920s. In 1926 Iowans held ninety-seven county fairs; three years later, this number had dropped to eighty-two. Even more ominous, most of the fairs still in operation barely turned a profit, and most of these would have lost money without their annual appropriation from the state government. Although county fairs struggled to survive, more than a million Iowans (in a state with a population of roughly two and one-half million) passed through their turnstiles in a typical year in the 1920s. Fair men attributed county fairs' difficulties to the growing availability of automobiles (by 1920, some three-quarters of Iowa farmers owned a car) and improved highways, which enabled farmers to travel more easily and bypass their county fair in favor of the much larger state fair. Many fair men believed that county fairs were failing because they offered too few entertainments and could not hope to compete with the lavish entertainments at the state fair.[40]

But some Iowans cited the dwindling number of county fairs as further evidence that the "drift to the cities" continued to undermine rural life. In 1930 *Wallaces' Farmer and Iowa Homestead* asked, "Will Our Local Fairs Survive?" The editorial charged that many fairs had slid toward bankruptcy because they had alienated farmers by permitting squalid sideshows to operate on the fairgrounds in a misguided attempt to increase revenues. County fairs, the magazine declared, were more closely connected to farmers than the state fair, and so had a special obligation to uphold their original agricultural mission. In 1921 one fair man urged his counterparts to band together and contract only with reputable booking agents, until the "junk" and "rotten" shows were "weeded out."[41]

In the 1920s, persistent worries about the future of rural life, compounded by troubling revelations of the wiles of the carnival business, led foes of amusements to become more vociferous in calling for entertainments to be eliminated from agricultural fairs. The crusade for "clean" fairs peaked in that decade, as critics of crooked and lurid sideshows at agricultural fairs redoubled their complaints about entertainments. Some of these would-be reformers, such as agriculturists and clergymen, had complained about the fair's amusements for decades. The movement to clean up carnivals was not led exclusively by moralists but also had influential champions within the outdoor entertainment industry, who sought to deflect, and even utilize, criticism of carnivals to weed out their own ranks and create a more reputable, streamlined, and prosperous industry. Afraid that critics would severely curtail or even ban carnivals at county and state fairs, showmen strove to clean up their industry, readily borrowing the rhetoric of preserving and improving country life to justify eliminating disreputable smaller shows in favor of larger carnival companies.[42]

The assault on crooked carnivals became a nationwide crusade in 1922, when *Country Gentleman*, an influential agricultural magazine published in Pennsylvania, launched an attack on midways by publishing the anonymous "Confessions of a Fair Faker," a scathing exposé of corruption in the carnival business that sparked widespread demands to clean up and strictly regulate outdoor shows. "My business," the anonymous faker admitted, "was to cheat farmers and their families." Overcome by remorse after a decade of swindling customers out of upward of $1,000 per week, he vowed "to take the cover off the rotten business and help clean it up." Schooled in the showman's art of luring patrons into games and sideshows, he now turned his well-honed powers of persuasion to exposing the crooked techniques of the carnival business. His heart-rending stories of poor farmers laying down their last hard-earned dollar in a futile attempt to win a rigged game and lurid tales of farm girls ruined or even killed after leaving home to join carnival companies shocked and titillated readers with their insider's lingo and melodramatic tone. Accompanying illustrations depicted "the Crooked Carnival" as a hydra-headed monster poised to destroy the countryside by spreading gambling, prostitution, and booze, and "the Crooked Midway" as a Pied Piper leading the unsuspecting young to their ruin.[43]

Country Gentleman credited "Confessions of a Fair Faker" with uncapping "a smoldering volcano of resentment against the unclean fair and carnival," and the magazine's mailbag bulged with letters from fair men and showmen applauding its stand against crooked carnies. "Confessions" sparked renewed scrutiny of the practices of carnival companies and inspired Don Moore, a fair man from Sioux City, to compile his own "Scrapbook of Fair Fakery," comprising newspaper items on crooked carnivals across the United States. Moore soon found himself "almost buried" beneath an avalanche of clippings, but he managed to dig himself out and publish excerpts from his "Scrapbook" in *Country Gentleman* the following year.[44]

Moore charged that the vaunted clean-up of the carnival industry in the early twentieth century had been little more than a shrewd public relations ploy and claimed that consolidation within the entertainment business had failed to eradicate rigged games and lurid shows. He counted 150 large carnival companies operating in the United States, 120 of which were "openly crooked or indecent," while the remainder "were decent only when they had to be." Far more worrisome was the untold number of "gilley" carnivals, which had no established itinerary, but pitched their tents wherever local authorities would permit them to work and flouted all standards of decency. "Except in those places where crooked carnivals have been prohibited by law or public sentiment," *Country Gentleman* reported, "there is not a corner in the whole map of the United States that is free from their leprous touch." Moore's scrapbook detailed the practices and the pervasiveness of carnival trickery and added to the groundswell of opposition to corrupt carnivals. In Iowa, the *Homestead* endorsed Moore's exposé and urged fair secretaries to "Cut Out the Fakes at Fairs," reminding them that "the morals of a community are worth very much more than all the money that can be taken in at the gates or realized from the sale of doubtful concessions."[45]

The outpouring of criticism of carnivals in the 1920s offered a reminder that some midwesterners, and even some fair men, still harbored considerable misgivings about the prominence of amusements at fairs. In the wake of this latest wave of criticism of the carnival business, showmen rushed to defend their industry and shore up its image. J. C. Simpson, former secretary of the Iowa state fair and now chairman of the World Amusement Service Association, a large entertainment company, insisted that amusements were essential, and not harmful, to fairs. The day had long since passed when amusements "were frowned on as wholly out of place," he wrote. "The average fair of today can look back to the time when it began adding a more varied program of amusements as the turning point in its struggle to keep going. It is no longer a question with the fair manager today as to whether amusements are necessary in his program of operation—but rather what form of amusements and to what extent they should be applied."[46]

When the Showmen's League of America, an association of carnival men, convened for its annual meeting in Chicago in 1924, Thomas J. Johnson, the league's legal counsel, warned showmen that they confronted a choice between self-regulation and extinction. The delegates responded by appointing Johnson as commissioner of the Showmen's Legislative Committee, granting him authority to expel crooked, irresponsible, or immoral showmen from the league and to notify fair men, mayors, sheriffs, and newspapers about carnivals that failed to earn the organization's stamp of approval. Just as Kennesaw Mountain Landis had been appointed to clean up and save professional baseball after the revelation that gamblers had rigged the 1919 World Series, and Will Hays had been hired to rescue Hollywood from the Fatty Arbuckle scandal, Johnson was entrusted to clean up the image of the carnival business.[47]

Showmen commonly deflected criticism of lewd shows and crooked games by accusing small, independent acts of tarnishing their industry. The Showmen's League discouraged fair managers from booking these entertainers, blaming "hop-scotch grifters," individual showmen who were not affiliated with large carnival companies, for carnivals' seedy reputation. Showmen and fair men joined forces in an effort to drive these smaller carnival companies and sideshows out of business, so that they would not expose the entire industry to charges of offending, cheating, or corrupting fairgoers. Fairs had become important venues for the outdoor entertainment business, and entertainments had become an indispensable aspect of the fair. Showmen and fair men resolved to regulate and clean up their own industry, before meddlesome legislators swept it away altogether.[48]

Billboard immediately pronounced this latest effort to purify the outdoor entertainment business a resounding success. The magazine reported that carnivals were no longer staffed by "the old-time grifting dregs of humanity" but by hard-working and highly skilled performers, mechanics, and carpenters. "They are human, intensely so," the magazine declared. "They have mothers, sisters, sweethearts, wives and homes. They often ask for newspaper clippings so that they can write home and show their mother the kind of show they are with." Only "bigoted fanatics who are opposed to the carnival form of amusement" could possibly find fault with the hard-working men and women who staffed modern carnival companies.[49] While nothing short of the complete exclusion of entertainments from fairgrounds would placate the most strident critics of amusements, showmen, fair men, and even some agricultural periodicals hailed the advent of "clean" carnivals. Even the *Homestead*, usually an unsparing critic of entertainments at fairs, proclaimed Johnson's effort to clean up fairs a success. "Never Were Fairs Better or Cleaner," declared the magazine in 1923, because fair men had finally learned that well-regulated, "clean" entertainments were preferable to tawdry sideshows. Two years later the magazine claimed that county fairs had rebounded from their financial doldrums, and it congratulated county and district fair men for cleaning up their fairs, "which seemed for a time to be headed for extinction."[50]

Although critics continued to complain about the evils of shows and games, the Midway had become every bit as much a fixture on the fairgrounds as the livestock barns. The entire fairgrounds was crowded and festive, but the Midway remained a world unto itself, humming with an energy and excitement of its own. A hodgepodge of tents and ramshackle structures, operated by transients who pulled up stakes and moved to a different fairgrounds every week, the Midway offered a stark contrast to the fair's livestock contests and 4-H exhibits. A reporter at the 1929 state fair described the Midway as a "gaudy circle of canvas, athrob with the cries of barkers, the sound of belly music, and the smell of cotton candy," and lined with freak shows, dancing girls, magicians, and daredevils—and thousands of customers, many of them farmers. Inside one tent, the crowd gaped while a performer "stabbed twenty-five swords through a young woman, poured oil on another young woman and set her afire, and shot a marked bullet through the chest of a third young woman." Spectators marveled at the Midway shows for a host of reasons, ranging from childlike wonder to morbid curiosity. Sideshows might not be edifying, the reported conceded, but, for the modest tuition of a quarter or fifty cents, they offered patrons a profound education in the ways of the world and the foibles of humanity.[51]

Canvas and Silver Screen

In the 1920s and 1930s, while fending off critics of shows and games, the outdoor amusement business also confronted stiff competition from upstarts within the entertainment industry. Outdoor showmen and fair men quickly recognized that the fair's real rivals were movies and radio, which made high-quality commercial entertainment, performed by famous actors and musicians, increasingly accessible to small town and rural Iowans. While critics still inveighed against entertainments, the real issue was no longer whether the fair would offer amusements—it would—but whether its attractions were exciting enough to compete with other entertainments for fairgoers' attention and money. Carnival companies, fair men, outdoor acts, and live performers of every stripe—carnivals, circuses, vaudeville, theater—all worried that they were being relegated to the margins of the show business by the astonishing spread of movies and radio in the 1920s and 1930s. In 1920 even the *Homestead*, usually a staunch champion of the superiority of farm life and critic of entertainments at the fair, wondered whether the *only* way to stem the exodus of young people to the city was "to transfer the city's amusements, as well as the city's comforts, to the farm." The *Homestead* predicted that rural electrification would soon enable each farmstead to be equipped with its own movie projector and that farm families would screen movies at home.[52]

Fairs and motion pictures had not always viewed one another as rivals. In the first decade of the twentieth century, movies remained a novelty and a rarity, especially for rural Americans, and traveling movie arcades did a brisk business on the state fair's Midway. When movie theaters and feature-length films supplanted arcades and nickelodeons in the 1910s and 1920s, theater managers and fair men became competitors, and feuding between theaters and outdoor acts dragged on for years, culminating in the late 1920s.

In 1926 *Billboard* magazine, which had gained its status as the "showmen's bible" by championing the outdoor entertainment industry, urged fair men and theater managers to "Live and Let Live," insisting that fairs and carnivals actually aided theaters by attracting potential moviegoers from the countryside into town. As movies continued to gain popularity, however, the magazine sounded a call to arms, declaring that "Outdoor Showmen Must Fight Their Oppressors." *Billboard* defended fairs, carnival companies, and vaudeville acts throughout the 1920s and 1930s, insisting that live performances delighted audiences in ways that movies could never match. The magazine gamely predicted that live entertainment would endure, while technological novelties such as radio, phonographs, and movies would prove faddish. Fairs and carnivals did survive, and even prospered, but the heyday of outdoor amusements had passed, as movies and radio became by far the most influential and lucrative branches of the show business. *Billboard* continued to cover outdoor amusements, but the magazine soon devoted far more attention to Hollywood, radio, and the recording industry. The outdoor entertainment business endured, but sawdust and canvas proved far less glamorous than the silver screen or the airwaves.[53]

The Night-Show Habit

Although fair men and carnival operators felt deeply threatened by the popularity of movies and radio, outdoor venues did offer some entertainments these new mass media could not duplicate. In the twentieth century, lavish nighttime spectacles, featuring enormous pyrotechnic displays, topped the fair's bill. For some three decades, the fair's amphitheater became the scene of nightly reenactments of history's most renowned battles and natural disasters. While critics grumbled that grandstand shows, like other entertainments, compromised the fair's educational mission, these spectacles provoked far less controversy than midways. Because they were staged at night, spectacles could not be accused of luring patrons away from the fair's agricultural exhibits, which were closed in the evening. As Secretary J. C. Simpson wrote to another fair man, "The night shows do not in any way detract from the useful and educational purposes of the fair." Spectacles might take some license with historical fact, but they were not crooked or fraudulent, and they did not entail the interaction with carnies or proximity to dancers or freaks common to games and sideshows. Although spectacles commonly featured dancing girls and scenes of outright debauchery, these temptations were kept at a safe remove from the grandstand. The popularity of historical spectacles embodied an important shift in American popular entertainment, as smaller shows that permitted and even encouraged the audience's participation were supplanted by shows in which they were mere spectators. Even *Breeder's Gazette*, a livestock journal frequently critical of midways, confessed that "it is easy to contract the night-show habit at these fairs."[54]

Billboard asserted in the 1920s that grandstand spectacles held the key to state fairs' success. When the fair's director of publicity, L. R. Fairall, polled patrons at the 1921 fair about what attracted them to the fairgrounds, 90 percent (!) replied that they came for the entertainments, especially the horse races, fireworks, and grandstand shows. As one showman wrote in 1926, rural Americans were no longer excited by fairs' small acts and sideshows because they had now seen better entertainments on the screen in their local theaters. Gigantic grandstand spectacles and fireworks displays, however, were one attraction that theaters could not duplicate.[55]

Gauging the meaning of these spectacles for their audience is not easy. Spectacle producers created eye-popping extravaganzas, which dazzled viewers with pageantry and pyrotechnics far surpassing anything that the local Chamber of Commerce could mount at the annual Fourth of July celebration. One woman, recounting her first trip to the state fair, admitted that

> I can't remember a single thing that I saw except the night show! "The Last Days of Pompeii" was being produced, and I sat on the edge of a narrow, hard old board, utterly lost in rapture—there was so much beauty, so much action, so much color, so much grandeur, in the setting against which the story moved to its spectacular and tragic conclusion with the eruption of Mount Vesuvius and the destruction of the city there at the water's edge!

On top of their already overwhelming scale and extravagance, spectacles also included a variety of vaudeville acts, adding to their festivity and perhaps distracting the audience from the show's sometimes conventional or flimsy plot. During "The Fall of Troy" in 1927, for example, a brass band serenaded the audience and "plump, graceful ladies in pink furbelows pranced along tight wires under the dazzling arcs" while the Greeks attacked the Trojans.[56]

Historical spectacles of a different sort also proved extremely popular in the early twentieth century, as progressive-era reformers staged locally produced pageants that recreated historical episodes in an effort to imbue ethnically diverse communities with a sense of shared heritage and faith in progress. Historical pageants produced by progressive, civic-minded organizations combined their promoters' nostalgic yearning for an idealized version of America's past and their optimistic hopes for the nation's future. In the 1920s, as the zeal for progressive reform waned, these civic pageants became more nostalgic and were staged to uphold "tradition" amid the dizzying pace of modern life and the bewildering changes transforming American culture. But commercial disaster spectacles, not civic-minded historical pageants, reigned as the most popular attractions at state fairs for some four decades, stretching from the 1890s to the 1930s. Their widespread, long-lived popularity suggests that many Americans did not altogether embrace the optimistic faith conveyed by Progressive Era historical pageants but harbored misgivings about the certainty of progress and its attendant costs.[57]

Disaster spectacles almost invariably culminated with the destruction of another, usually non-Western, civilization, and they frequently bespoke white Americans' disdain for other cultures as the United States emerged as the world's most powerful nation. After the set had been reduced to smoldering rubble, the pyrotechnic grand finale often included a number of patriotic set pieces (images outlined by fireworks mounted on wooden scaffolds) depicting the American flag and other national icons.[58] But spectacles could be racist, xenophobic, nationalistic, and simultaneously suggest anxiety about the United States' newfound power and prosperity. Reenactments of the sudden, inescapable demise of thriving civilizations could hardly fail to remind spectators that even the world's greatest empires ultimately succumbed to decay, defeat, or destruction. The spectacles' cataclysmic finales conveyed a simple moral: progress, power, prosperity, and pleasure sometimes exacted a high price and, ultimately, culminated in disaster.

Performed before an audience of thousands without electric amplification, these shows did not contain subtle dialogue or complicated plots. In order to be intelligible to their viewers, spectacles had to depict broad themes, clear-cut conflicts, and unambiguous outcomes.

Pyrotechnic companies typically produced only one or two spectacles each year, and they took their show on the road to fairgrounds around the nation. Although these spectacles were not written specifically for a midwestern or rural audience, fairgoers likely drew their own particular lessons from them, lessons invariably shaped by the tensions that swirled around the fair as a rural, agricultural state adjusted to a rapidly urbanizing nation. The civilizations annihilated before the fair's amphitheater were usually great cities bustling with festivity in their last moments before sudden, unforeseen disaster. Disaster spectacles preached the lesson that the frenetic pace and frivolity of urban life were dangerous, even wicked, and that they would ultimately provoke retribution. In short, commercial grandstand spectacles, immensely popular at state fairs across America from the 1890s into the 1930s, presented a version of history at odds with the soothing bromides of civic pageants or the paeans to progress annually invoked by the fairs' organizers.

The state fair staged its first gargantuan disaster spectacle in 1902, booking the Pain Pyrotechnic Company's reenactment of the eruption of Mt. Vesuvius. The pageant's popularity with fairgoers led the fair's secretary to book the company's pageant "Ancient Rome," which depicted the fire that consumed the city during the reign of the licentious Emperor Nero, for the 1903 fair. For the next thirty years, gigantic pyrotechnic spectacles headlined the fair's bill of attractions. Apart from their common motif—virtually every spectacle culminated in utter destruction, which leveled the set and everything on it—spectacles could be categorized into a few genres: battles, natural disasters, and fantasies.[59]

Spectacles sometimes enabled the audience to witness reenactments of famous battles and other events from recent history. In 1905, for instance, the fair's managers booked a re-creation of the Japanese attack on the Russian naval base at Port Arthur the previous year. In 1914, shortly after the completion of the Panama Canal, the fair's nighttime entertainment featured "The Opening of the Panama Canal and Uncle Sam's Reception to the World." Billed as "a realistic, scenic production," the spectacle boasted three hundred cast members on a four-hundred-foot stage. After a lavish ceremony to celebrate the canal's official opening, Panama suddenly became the scene of "a gorgeous pandemonium of cannonading and multi-colored fire," as ships and aircraft circling overhead engaged in battle. The eruption of a nearby volcano (nowhere to be found in the topography of the real Panama) supplemented the aerial and naval bombardments, destroying all the buildings lining the waterway and the ships in it. When the guns fell silent and the smoke finally cleared, "the whole festive scene had been shot, blown, and burned to fragments."[60]

From 1915 to 1920, the all-too-real disaster in Europe provided themes for the fair's spectacles, as producers staged recreations, however loose, of major battles. After the United States entered the war in 1917, virtually all of the fair's activities, from home canning to Midway shows, were imbued with patriotic and military themes. Pyrotechnicians and fair managers alike promised the fair's patrons that the grandstand spectacles would afford them an understanding of the war's horrors that could otherwise be gained only by witnessing the battles in person. The fair's advertisements declared that newspaper accounts of the war offered a feeble substitute for seeing, because "pictures appeal with peculiar force and significance." The fair's grandstand shows for 1917 and 1918, "Modern Warfare, or an Attack on the Trenches" and "World's War," were jingoistic productions designed to provoke revulsion at German attacks on French and Belgian villages, and they concluded with rousing patriotic music and pyrotechnic displays of the American flag, Statue of Liberty, and other national icons.[61]

Historical events also inspired several grandstand spectacles, although these events were almost invariably embellished to thrill the audience. Many of these historical spectacles depicted the United States' military conquests. In 1913 fairgoers were treated to a reenactment of General Winfield Scott's capture of Mexico City in 1847, supplemented by a volcanic eruption of nearby Mt. Popocatapetl, which exploded just as the battle for the Mexican capital reached its climax. As the spectacle opened, Mexican citizens joyously celebrated a feast day by engaging in a variety of native games and dances, only to be annihilated by the combined force of American troops and molten lava. In 1928 the fair presented the "Battle of Manila," which commemorated the thirtieth anniversary of Admiral Dewey's defeat of the Spanish fleet in the Philippines and America's rise to global prominence.[62]

Even more distant historical battles and disasters also proved popular with fairgoers. The fair presented "The Fall of Rome," depicting the city's burning under Emperor Nero, in 1903 and again as "Rome Under Nero" in 1925. Advertisements for the 1925 pageant boasted that it was surpassed "only by the original burning of Rome under Nero," and that "every effort has been made to keep the presentation of the production accurate both as to realistic scenic effects and to historical truth." In

1927 "The Fall of Troy" filled the fair's stage. Taking a bit of license with Homer's epic tale, the production required "an army of pyrotechnicians" and promised "an entire city blown up before your eyes." In 1921 the fair staged "Montezuma," which featured a battle between the great chief's Aztecs and Cortez's conquistadors. As usual, the spectacle's creators added natural disasters to complete the annihilation of the Aztecs' civilization, embellishing the Spaniards' conquest with a volcano and an earthquake.[63]

Natural disasters, like battles, lent themselves nicely to the pyrotechnicians' art and penchant for destruction. Vesuvius erupted repeatedly, burying Pompeii in 1902, 1907, 1911, 1916, and 1929, until few Iowans had not vicariously witnessed history's most famous natural catastrophe at least once. As depicted before the fair's grandstand, however, the volcano's eruption seemed not so much a geological phenomenon as an outburst of moral retribution against excessive festivity. The 1916 production of "The Last Days of Pompeii," for example, opened with a scene of Pompeii "in all its splendor on a fete day, with its crowded streets, gladiatorial combats, races, games, Egyptian dances and other forms of merrymaking," until Vesuvius erupts, "burying the city with its profligate people." At least a few fairgoers must have perceived that the games and revelry in Pompeii resembled the festivity all around them on the fairgrounds.[64]

In 1929 the fair celebrated its seventy-fifth anniversary by erecting yet another cardboard Vesuvius and staging a new version of "Pompeii," billed as the "greatest, most appalling fireworks spectacle of all time," not to mention the most frequently produced. Recent natural disasters also became the stuff of spectacle. At the 1924 fair, spectators witnessed "Tokyo Through Quake and Fire," which depicted the disastrous earthquake that had struck the Japanese city a year before.[65]

Pyrotechnicians sometimes devised more fantastic spectacles, in which they could indulge their imaginations untrammeled by any need to depict actual events. In 1909 spectators filled the fair's new grandstand to witness the Pain Company's futuristic "Battle in the Clouds." In the year 2000, while citizens of Centaerial, a utopian "city of science," celebrate the anniversary of a decisive aerial victory over their rivals, their city is suddenly attacked by enemy airships. Troops loyal to King Pyro, "ruler of the world of science," fight valiantly in self-defense but are machine-gunned by the invading aircraft, which proceed to reduce the city to a pyrotechnic "fire of all colors." According to the fair's advertisements, the production offered viewers a glimpse of "the possibilities of future warfare in the skies" utilizing airplanes, machine guns, and bombs. But rapid advances in military aviation would soon outstrip pyrotechnicians' imagination, making the spectacle's aerial bombardment look like a hokey vaudeville stunt.[66]

In the 1920s, after several years of spectacles re-creating horrific battles from World War I, spectacle organizers shied away from recent events and created spectacles about utterly mythical disasters. When pyrotechnicians let their imaginations run riot, they preferred to wreak havoc on Asian and Middle Eastern cities. In 1922, for instance, the fair presented "Mystic China," which purported to show "in a weirdly fantastical and mysterious manner the curious and mystic customs of the people of the 'Flowery Kingdom,' whose past is shrouded in obscurity." After celebrating several feasts and festivals, the Chinese watch helplessly as their city is invaded and sacked by marauding Mongols. The following year, the grandstand stage became "India," a spectacle that included live elephants, a suttee, and a Hindu mutiny. British troops suppressed the throngs of rebellious Hindus and destroyed Delhi, after which a pyrotechnic blaze swept across the entire stage, consuming whatever remained of the devastated city. In 1928 the fair presented "A Night in Baghdad," which began with scenes of revelry and concluded with a titanic struggle between good and evil genies atop the set's eighty-foot mountain, culminating in the entire peak being "destroyed in an earth-shaking clash of the opposing forces."[67]

Commercial spectacles conveyed a worldview that was far from optimistic, suggesting that history is punctuated by periodic catastrophes, and that power and pleasure-seeking often led to ruin. In 1930 the Thearle-Duffield Fireworks Company audaciously attempted to chronicle "the entire course of world history" in "The Awakening," a spectacle that declared unequivocally that "progress" had gone too far. The gargantuan spectacle opened with the "awakening" of both life and evil in the Garden of Eden and portrayed, in turn, orgies and persecution of Christians under the Roman Emperor Nero, Columbus's discovery of the New World, the American Revolution, and the Civil War. The awakening of a new evil, "the spirit of jazz, unrest and laxity of living," marked the next epochal event in this peculiar timeline. Jazz dethroned the spirit of Beauty, provoking Americans to "throw themselves into an orgy of abandonment," dancing, and drinking, until a pyrotechnic "outburst of the heavens" squelched the jazz orgy.[68]

Disaster spectacles were often unabashedly nationalistic, xenophobic, and racist, and they pandered to white Americans' contempt for other cultures. Yet it would be too simplistic to view these spectacles as one-dimensional jingoistic tableaux of American or Anglo-Saxon superiority. In an era in which many Americans were anxious about the vast changes rapidly remaking their nation, and in which midwesterners were particularly concerned about the advent of an urbanized, consumer-oriented society, depictions of revelry, corruption, and the sudden, violent demise of vibrant civilizations not only allowed spectators to gape at others' misfortune but implicitly suggested to at least some fairgoers that their own society might be hurtling toward a similar fate.

But after a run of some thirty years, the popularity of grandstand spectacles plummeted as the economic disaster of the Depression made staged catastrophes far less appealing in the 1930s. Amid the economic troubles and political uncertainty of the Depression decade, Americans once again sought reassurance in their nation's past and searched for omens that prosperity and stability would return in the future. They were hardly disposed to pay for extravagant spectacles chronicling troubled times and culminating in utter disaster. The growing popularity of movies, which rebounded from the shock of the Depression to flourish in the 1930s, also contributed greatly to the waning of grandstand spectacles. Now that most fairgoers could see and hear historical and dramatic films in their local theater, gigantic stage shows seemed downright slow-paced and ponderous, and these lumbering dinosaurs, which had ruled the fair's entertainments for decades, quickly succumbed to extinction.

Keep the Fairs Going

The Great Depression, a calamity that dwarfed even the most apocalyptic grandstand spectacle, forced the fair to cut back on expenses and led its secretary to tout the exhibition's role in restoring economic prosperity. After the stock market crashed, showmen and fair men, who spoke only the language of boosterism and hype, insisted that Iowans cherished their annual fair so deeply that the Depression would not keep them away during fair week. The annual carnival proved unable to defy the economic downturn, however, and its attendance and receipts, like virtually every other indicator of economic well-being, plummeted. The ravages of the Great Depression left the fair's secretary unwilling and unable to book expensive grandstand shows, forcing him to rely instead on more modest acts. In 1931 and 1932 the fair replaced its enormous grandstand spectacles with variety shows, which were much cheaper to book, and attracted fairgoers with songs and comic sketches designed to appeal to an audience familiar with radio programs and movies.[69]

Confronted with the daunting task of trying to lure Iowans to the fair during the Great Depression, fair men simultaneously extolled the fair's agricultural roots and ballyhooed its entertainments, billing the 1932 fair simultaneously as an "old fashioned" fair and as a "Fair of Thrills," which promised fairgoers "a thrill a minute." By far the biggest of these thrills—and perhaps the most sensational entertainment in the fair's history—was a different sort of disaster spectacle, a head-on collision of two obsolete steam locomotives in front of the fair's grandstand. Amid the economic wreck of the Depression, the wanton destruction of two railroad engines seemed both a colossal waste and eerily irresistible.[70]

"Thrill Day" dawned auspiciously, as patrons began streaming into the grounds in the early morning. *The Des Moines Register* estimated the grandstand crowd at forty-five thousand, but the fair's secretary doubled that figure. Anticipation built as the huge crowd was treated to a few warm-up acts, including automobile crashes and auto-train collisions—billed as "safety demonstrations," which would instruct spectators about the importance of exercising caution at railroad crossings(!). The headliners, hundred-ton engines prominently labeled "Hoover" and "Roosevelt," faced one another like jousting knights on a track laid in front of the fair's grandstand. To ensure that the collision of the hundred-ton engines would be sufficiently spectacular and destructive, the track was inclined at each end to enable the locomotives to accelerate to a speed of fifty miles per hour in their brief run, while the engines' cowcatchers were primed with dynamite and their coaches contained open buckets of gasoline. Hoover and Roosevelt hurtled down the track right on schedule, slamming into one another directly in front of the grandstand, producing an ear-splitting crash and explosion that reduced both engines to piles of twisted metal. Although the obsolete locomotives were already destined for the scrap heap, their gratuitous destruction, along with promoter Joseph Connolly's whopping fee of $40,000, prompted some grumbling about profligacy at the depth of the Depression, grumbling that grew louder after the Fair Board announced that the 1932 fair had lost $67,000.[71]

The Depression not only forced fair men to scramble to book entertainments that would lure crowds to the fair but also prompted them to dust off and reuse well-worn rhetoric about the fair's "true" agricultural mission. "The Iowa State Fair feels a stern responsibility in times such as these," the 1931 premium list declared, adding that the fair would certainly provide diversion to its visitors but would be "primarily as a place to go to school." The following year, the Fair Board announced that its goal was nothing less than "the rehabilitation of agriculture, and the restoration of the farmer to his rightful place in the economic world," promising that the exhibition would be "a farmer's fair" and "an 'old fashioned' State Fair," one that aimed above all else to educate its visitors about agriculture and home economics. In 1933 the board adopted "Back to Fundamentals" as the fair's slogan, declaring that "the time has come for agriculture to return to the proven principles which gave our farm regions their first development and prosperity." Hard times led midwesterners, like other Americans, to search the nation's history for the "proven principles" that had once made their agricultural economy prosperous and lent the fair its sense of high purpose, and impelled fair men to sound strangely like old-time agriculturists. Even *Billboard*, which had always maintained that fairs were in the show business, sometimes borrowed phrases straight out of agricultural periodicals as it championed the interests of the outdoor entertainment industry during the Depression. *Billboard* implored legislators and fair men not to cancel their fairs for financial reasons. "Keep the Fairs Going," the magazine urged, reminding its readers "how elemental our fairs are in our whole program of agricultural progress."[72] A few months later, the magazine ranked fairs as more important than public schools as educational institutions for farm families.[73]

Fair men were well versed in agrarian rhetoric, and they employed it liberally in the state fair's advertisements and press releases, but the state fair was by now indisputably in the show business. Fair men and showmen worried that farmers would be unable to afford their annual trip to the fair, and that the Depression would decimate the already beleaguered outdoor entertainment industry, which simultaneously had to fend off moralists and compete with movies, radio, and other entertainments for whatever money Americans had to spend on entertainment.

Billboard worried that fair secretaries would economize by slashing amusement budgets and that agriculturists and cranks would seize on the Depression as a pretext to eliminate fairs' amusements entirely. According to the magazine, fair men should resist the obvious temptation to spend less on entertainments, because the only way to keep fairs profitable was to offer more and better attractions. Only professional showmen, not agriculturists, could keep the fair afloat during hard times. "After all," *Billboard* reasoned, "you wouldn't call a lawyer to prescribe for your sick child, nor a doctor to handle your legal affairs." "It is all right to argue that it [the fair] is educational," the magazine declared in 1930, but "a fair is a show and unless it is sold and promoted in the manner of a show its utmost cannot be realized." Even at the depth of the Depression, fair secretaries knew that a fair without entertainments would inevitably fail. Although fair men resorted to agrarian rhetoric to promote fairs in the 1930s, the Depression hardly caused them to pine for a return to a bygone, largely mythical era in which fairs were strictly agricultural exhibitions.[74]

Although industrialization had transformed the American economy in the late nineteenth and early twentieth centuries, Jeffersonian rhetoric about the superiority of agriculture remained influential. The Depression, coupled with misgivings about the spread of a consumer-oriented culture that emanated from the nation's urban centers, rather than from the rural Midwest, prompted some writers and artists in the 1930s to stake their claim for an indigenous regional culture, one predicated on agriculture. Writers and painters created iconic images of midwestern farm life, insisting that the combined forces of urbanization, industrialization, consumerism, and Depression had not uprooted the nation's agrarian heritage. Their paeans to farm life won admirers not only in Iowa and the Midwest but across the nation. Ironically, however, their effort to establish an authentic midwestern culture attested to the extent to which the Midwest was not an autonomous region but was integrated into a larger national culture and consumer economy.

5) Agricultural Lag

Writing in the *American Mercury* in 1926, Iowan Ruth Suckow diagnosed a collective psychological malady afflicting her native state. Iowa had acquired "a timid, fidgety, hesitant state of mind" with regard to cultural and intellectual matters, the result of decades of dependence upon New England for guidance in religion, learning, and the arts. Iowans' cultural inferiority complex sprang from deep historical roots, according to Suckow. Iowa's original settlers had come "with the belief that they were leaving culture behind." They had come to the Midwest to acquire farms and make money, and they lacked confidence that they could create their own indigenous culture in their new environment. The experience of settling and living in the Midwest, most Iowans assumed, could not inspire artistic creativity. Iowa might be fertile and productive, but "Culture, art, beauty were fixed in certain places." Judged by the standards of Boston or New York, the Midwest could never be anything but uncivilized. But Suckow detected some hopeful signs that Iowans were at last beginning to shake off their feelings of inferiority toward the East and develop "a native culture." The sheer expanse of the West and its settlement by diverse ethnic groups had finally snapped New England's cultural hegemony, precipitating a tectonic shift, a "general break-up of culture," which had enabled Iowans to discuss and even create literary and artistic works. Suckow found it difficult to predict exactly what shape this emerging culture might take: "All the elements, old and new, are jumbled up together until it seems impossible to guess what can be fished out of the middle." She likened this emerging midwestern culture to a geological formation that could best be envisioned in cross-section. The top layer of Iowans' chronic self-deprecation was quickly eroding; beneath it lay the outmoded cultural ideals of New England and Europe, the boosterism of businessmen, and "the Main Street element of small town hardness, dreariness and tense material ambition." Underlying all these layers were the farmers, "still the very soil and bedrock of our native culture." Suckow considered rural life the solid "bedrock" of Iowa's culture, without which "what we call culture in Iowa would be as insipid as cambric tea."[1]

Suckow's was only the latest voice in a conversation stretching back to the settlement of the Midwest. Could a society overwhelmingly reliant on agriculture eventually give rise to a distinct civilization, one capable of producing its own literature and art? Suckow espoused an influential strand of midwestern thought in the early twentieth century, which insisted that any authentic midwestern culture would invariably be built atop the region's agricultural abundance. Importing culture from the East or from abroad would never produce anything but pale imitation; an indigenous midwestern culture would have to spring from the same source as the region's wealth—from agriculture itself.

Predicting the advent of a distinctive midwestern culture in the 1920s was not a bad bet. An impressive group of midwestern writers, including Willa Cather, Laura Ingalls Wilder, Sinclair Lewis, Sherwood Anderson, and Iowans Hamlin Garland and Herbert Quick, had earned critical acclaim. In a decade during which tensions between rural and urban America ran unusually high, F. Scott Fitzgerald's tale of greed, hedonism, and duplicity in the East, *The Great Gatsby*, concluded with a wistful over-the-shoulder glance toward the Midwest, "where the dark green fields of the republic rolled on under the night," and where Americans were not corrupted by materialism, self-indulgence, and dishonesty.[2]

Just at the moment when the United States plunged from the dizzying prosperity of the 1920s into the Great Depression, Phil Stong, an aspiring novelist from southeastern Iowa, and a remarkable group of self-proclaimed "regionalist" painters staked bold claims for a distinctly midwestern culture. Fittingly, the state fair, Iowa's central cultural institution, furnished both a subject and a venue for their stories and paintings. As the Great Depression tested Americans' faith in their society, regionalist writers and painters reassured them that the traditional virtues of rural life still thrived in the nation's heartland. Ironically, regionalists' attempt to create a vibrant midwestern culture registered an eleventh-hour protest against the national economy and mass culture in which the region was already inextricably enmeshed.

The Fair in Fiction

The most popular state fair of the 1930s was not to be found on a fairgrounds but between hard covers and in theaters. In 1932 Iowa native Phil Stong published his first novel, *State Fair*, which became the basis for the successful Fox Studio film released the following year. The state fair, whose popularity had been challenged by the spread of motion pictures and other mass entertainments, now became the backdrop for one. While farmers and fairs around the country struggled to weather the Depression, Stong and Fox scored a runaway hit with their idyllic depiction of farm life, in which hard times were nowhere to be seen. Critics charged that both the novel and the movie cavalierly overlooked the economic troubles besetting American farmers in the 1930s. But *State Fair* did more than merely ignore the Depression: it depicted rural Iowa as a land outside of time, virtually impervious to the effects of urbanization, mass culture, or the ravaged economy.[3]

Although it recounts the annual pilgrimage of the Frake family from their southeastern Iowa farm to Des Moines, *State Fair* conspicuously failed to evoke the language or concerns of farm life in the 1930s. Instead, Stong invented a genial, well-scrubbed midwestern farm family, the Frakes, who were appealing to Americans everywhere. The extraordinary success of the novel and film *State Fair* attested to the triumph of mass culture, rather than the creation of an indigenous midwestern culture. Stong packaged Iowa as though it were a wholesome, trusted name-brand product, to be marketed coast to coast to satisfy Americans' enormous demand for images of bucolic rural life, untainted by urbanization or by the Great Depression.

At the same moment *State Fair* became a sensation, Grant Wood and a group of self-proclaimed "regionalist" artists created their own image of the Midwest by painting rural landscapes and scenes of everyday life. Because the Iowa State Fair's Art Salon was the only widely viewed and influential art gallery in the state, regionalists entered their work in the fair's contest, where Iowans variously found the paintings appealing and unflattering. Despite considerable opposition, regionalist painters carried off nearly all of the Art Salon's blue ribbons in the 1930s. At the same time regionalists' influence was spreading rapidly beyond Iowa and the Midwest, as regionalism became both enormously influential and controversial across the United States, alternately hailed for creating iconic images of Americana and assailed as homegrown fascist art.

Phil Stong and Grant Wood both created images of Iowa that remain indelible to this day. Yet both the novelist and the painter succeeded in part because they depicted an appealing image of Iowa recognizable to Americans everywhere, one designed to assure viewers and readers that America's agrarian foundation remained solid despite urbanization, industrialization, and the Great Depression. Although Stong and Wood proclaimed their determination to depict an authentic version of Iowa life, they offered instead an idealized vision of the rural Iowa as a repository of virtue, modesty, and common sense.

Iowa Roots

Phil Stong's Iowa roots ran deep. His maternal grandfather, George C. Duffield, arrived in the new territory as a boy in the 1830s and became one of the young state's most prominent settlers.[4] Phil's father, Ben, owned a small dry goods store in Keosauqua. He sent his son to Drake University in Des Moines to prepare for a career in business or law, but Phil soon devised other plans: "I am going to be a writer....The fire is in my mind, and it will burn a way for me in the world, sometime."[5] After graduating from Drake in 1919, Stong returned to Des Moines in 1924 and landed a job writing for *The Des Moines Register*, covering dozens of stories, including the state fair.[6] Dissatisfied with journalism, Stong yearned to become a novelist. In 1925 he moved to New York, where he held a series of newspaper and advertising jobs and churned out novels and short stories in his spare hours.[7]

By 1931 Stong had written twelve novels—none of which had been published. Shortly after he began pecking out his thirteenth, his literary agent called to say that publishers were avidly searching for "a Sinclair Lewis story more humorous and fairer to small town people than *Main Street*." Stong suggested a story, "State Fair," about a farm family's experiences during fair week. Encouraged by his agent's enthusiastic response, Stong hurriedly cranked out *State Fair*, which was published by the Century Company of Philadelphia in 1932.[8]

Stong predicted modest success for the novel. "I don't think it has the stuff for a resounding success, but I think it will sell 10,000 copies, finally, get fairly good reviews and give me a small entrance to the literary field."[9] But *State Fair* rapidly exceeded Stong's expectations, climbing onto bestseller lists in New York, Chicago, St. Louis, Kansas City, and other cities. When the Literary Guild named the novel a Guild selection, sales skyrocketed. As Stong excitedly noted, he would now surely land a movie deal for *State Fair*, and publishers would vie to purchase his next manuscript. Most important, "it means that I'll be taken seriously by critics and readers right from the jump, instead of as a fiction hack; and it means that I can kick this #$§$#&¶§$# [advertising] business in the pants and never do anything or go anywhere I don't want to again."[10] Elated by the success of *State Fair*, Stong now began to look forward to a movie deal, which included a job as a screenwriter "at some kind of crazy salary." In early June, he wired his parents triumphantly,

> State Fair sold fox fifteen thousand dollars twelve weeks Hollywood at three hundred dollars first six weeks three hundred fifty after stop option to be called prior to August plans for coming home depend on these arrangements love to all
>
> Phil[11]

In the depths of the Depression, Phil Stong had struck it rich.[12]

State Fair

Stong's novel begins in a country store on a languid Saturday evening, as Abel and Melissa Frake complete their weekly shopping trip into Brunswick. Overseeing the store, and the entire story, is the Storekeeper, a genial cynic who wryly observes humanity's foibles from his vantage behind the cash register. In the Storekeeper's mildly jaded worldview, the fate of human beings is governed by meddlesome deities known only as "Them," who "ordain all things for the worst—but more mischievously than tragically."[13]

The Frakes eagerly anticipate the upcoming state fair, where Abel confidently expects his magnificent Hampshire boar, Blue Boy, to capture the blue ribbon. The Storekeeper taunts Abel, warning him that if Blue Boy is judged the state's finest hog, "They" will ensure that Abel gets his comeuppance by suffering some calamity. "Don't let your hog get too good," warns the Storekeeper, to which Abel grumbles, "Don't say anything against that hog. I've got faith in my hog. I believe in my hog."[14] Abel wagers the Storekeeper five dollars that Blue Boy *will* win the grand prize, and that no disaster will befall the Frakes, who will "all have a good time and [be] better off for it when the whole Fair is over." The Storekeeper tacks on a side bet: "And I bet you that if I lose, something will've happened that we don't know about but that will be worse than anything you can think of. They're tricky." The two shake hands, and the novel's mainspring is set.[15]

Meanwhile, the Frakes' teenage children, Wayne and Margy, both struggle with the pangs and frustrations of romance. Wayne frets that his girlfriend, Eleanor, has become too sophisticated during her first year away at college; Margy recoils upon discovering that her devoted but unexciting boyfriend, Harry Ware, has already planned out her entire life as Mrs. Harry Ware, farmwife and mother of a brood of strapping farmboys. Dissatisfied with their hometown sweethearts, the younger Frakes eagerly look forward to a week at the fair, with its prospect of new sights, new faces, perhaps even new romance.[16]

The Frakes' journey to the fair is no ordinary vacation but a collective escape from the daily labors and isolation of farm life. Abel, Melissa, Wayne, and Margy simultaneously experience "almost an eerie feeling" as they climb into the family pickup to drive to the fair. As they rumble through the gate of the Frake farmstead, the familiar road into town "suddenly seemed strange because at this time it led to strange and romantic places." Loosed from the familiar moorings of their farm and hometown, the Frakes suddenly realize "that life went on, far outside their consciousnesses, in many places and at all times."[17]

A few hours later the Frake pickup rolls through the gates of the state fairgrounds and the family pitches camp. Abel and Melissa, predictably, meet with success during fair week: Blue Boy is crowned the state's finest hog and Melissa's pickles are awarded the blue ribbon. While Abel and Melissa anxiously await the judges' decisions in the Livestock Pavilion and Exhibition Building, Wayne and Margy head straight for the pleasures of the Midway. Wayne has spent an entire year itching for revenge on a crooked carnival barker who cheated him at last year's fair, but he soon discovers a game more fun than ring toss, when he meets Emily, a spunky young woman utterly unlike the girls in Brunswick. Emily had neither a mother nor a home, but had grown up living in hotels and traveling to fairs and horse shows across the country with her father, an incorrigible gambler and womanizer. Her rootless, hedonistic life runs counter to Wayne's work ethic: "It's so much fun to do things you want to do, that don't hurt you or anybody, that I nearly always do them," she explains. "I've lived on a farm my entire life," Wayne replies, "and I've never seen anybody the least bit like you."[18]

Meanwhile, Margy also ventures to the Midway to ride the roller coaster. Boarding the contraption, she finds herself seated next to Pat Gilbert, a young reporter for *The Des Moines Register*. Like Emily, Pat is cosmopolitan and hedonistic, and has written for newspapers across the United States. For Pat, as for Emily, "good fun was its own excuse." Riding the roller coaster becomes Stong's metaphor for sexual intercourse, and his visceral descriptions of its slow rising, the riders' anticipation, and the coaster's exhilarating plunge, along with Margy's rapidly growing eagerness to hop aboard the contraption again and again, evoke the couple's whirlwind relationship.[19] Both Wayne and Margy lose their virginity within days of arriving at the fair, as Emily seduces Wayne in a downtown hotel and Pat makes love to Margy in a grove near the Frakes' campsite. Critics of the novel found Wayne's and Margy's sexual adventures implausible or even immoral and insisted that the growing toleration of youthful sexuality in the 1920s was exclusively an urban phenomenon, which had not affected teenagers across rural America.

As fair week draws to a close, the young couples worry that their new romances will prove as fleeting as the fair. Pat sweet-talks Margy, whispering that the end of the fair need not spell the end of their romance. When he proposes to her, she rejects him, informing him that "if you marry me you'll tie yourself down to one place, that you don't like, and one woman, all your life." He vows to "manage myself . . . like a Frake," but excitedly predicts that his career will soon catapult the couple out of Brunswick to Manhattan. Certain that she could never adapt to the pace or sophistication of life in New York, Margy informs Pat her destiny is to become Mrs. Harry Ware of Brunswick. Although she does not love Harry, "I love—his kind of life. I'd be—somebody—back there." The couple stroll to the secluded grove near the fair's campground, where they make love for the last time while the distant clatter of carnies striking tents signals the fair's imminent end. Both Pat and Margy reluctantly concede that their lives are incompatible. Only at the fair could an innocent farm girl and a streetwise reporter find romance together, and then only briefly. The following morning, after a week of hectic days and sleepless nights, Margy's thought begin to drift from the unreality of the fair back toward her familiar routine in Brunswick. She wonders aloud, "I love you, Pat, but sometimes you seem like something—I'll wake up from. But I'll write—It's so strange—if it had been back home—"[20]

Wayne desperately pleads with Emily to become Mrs. Wayne Frake. She has other plans, informing him that she intends to blow all of her racetrack winnings on one night on the town and then "bid you a fond farewell." Realizing that Emily has no hankering to become a farmwife, Wayne assures her that hired hands would spare her from manual labor, and that the tedium and isolation of rural life would be alleviated by jumping into the pickup and driving to Fairfield, Farmview, even Ottumwa. Emily explains to Wayne that their lives are incompatible: "Can't you see that even though we could marry and be happy, our lives can't marry and be happy?" Reluctantly, he concedes that she is right: "I was raised to run a farm—and you weren't—and we couldn't ever reach middle ground." Wayne and Emily, Margy and Pat—rural and urban youths might cross paths at the fair, but in Stong's novel farm people would invariably return home, while city youth zoomed off to destinations unknown.[21]

On Saturday evening, Abel Frake, the human embodiment of "a kind of Destiny poised over what had been Fair Week," clambers behind the wheel of his pickup to head home. The fairgrounds has fallen dark, empty, and silent "where all the sparkle of the other nights had been." The trip to the fair had transported the Frakes to a world in which they temporarily slipped free from the constraints of farm life, while the drive home returns them to familiar surroundings and their workaday routine. Exhausted and despondent, Margy and Wayne slump in their seats. Melissa, oblivious to the reason for their sullen moods, attributes them to the fair's end. "My goodness," she remarks innocently, "you don't seem to me like the same youngsters came up to the Fair with me."[22]

Upon returning home, Margy wakes drowsily to the eerie sensation "that there had been no Fair, no change, no Pat." The following afternoon resembles any other Sunday on the Frake farmstead. Wayne and Eleanor are reunited, as are Margy and Harry, and the Frakes enjoy their traditional Sunday dinner. The Storekeeper drops by to survey the Frakes' condition upon their return from the fair and collect his five bucks. When he asks the younger Frakes whether they enjoyed the fair, they shrug, but almost imperceptibly, "between Wayne and the Storekeeper and Margy and the Storekeeper there passed a secret glance of understanding." Assured that he has in fact won his bet with Abel, the Storekeeper magnanimously forfeits his wager, since Abel and Melissa would be shocked to learn that something truly "worse than anything you can think of" had befallen their children during fair week.[23]

Ironically, Abel had also wagered correctly that the Frakes would "all have a good time and [be] better off for it when the whole Fair is over." Abel would have flown into a rage and Melissa would have fainted if they learned of their children's sexual liaisons during the fair, yet Wayne's and Margy's sudden coming-of-age actually renders them more content with farm life. Neither Wayne nor Margy suffer any lasting emotional scars from their weeklong flings, and neither succumbs to the temptation to forsake the farm for the city. In response to Americans' hand-wringing about the "drift to the cities," *State Fair* suggested that the urban life and mass culture would not lure young people away from the farm. Wayne and Margy are as intelligent and adventurous as city youth, but they do not crave the city's excitements enough to abandon the farm. Stong insists that farmers and urbanites remain, at heart, fundamentally different. The Frakes, and all farm families, are deeply rooted in the land, attached to their home, and devoted to their forebears. Their virtually timeless existence, governed by the recurring cycle of planting and harvesting, spans generations without significant change. Urbanites, by contrast, lead rootless lives, chasing new experiences and changes of scenery. Only in the enchanted arena of the state fair, and then only fleetingly, is the chasm between rural and urban Americans bridged. When the fair ends, farm folk are happy to return home to the farm.

Mr. Stong's Dreamy Iowa

Stong's prediction that *State Fair* would "get fairly good reviews" proved half right. Critics divided into two camps: those who considered the novel an enjoyable diversion and those who scoffed at its upbeat depiction of rural life at the depth of the Great Depression. To Stong's annoyance, his former employer, *The Des Moines Register*, gave the book a chilly reception. Reviewer Donald Murphy applauded Stong's attempt to write "a moderately cheerful book" about midwestern farm life, one that countered the bleak tales of realist writers such as Hamlin Garland and Erskine Caldwell. But Murphy detested the novel's scandalous love scenes, which, he assured his readers, "are far from typical" at the state fair. Stong's pleasant depiction of rural life provoked several reviewers to dismiss *State Fair* as little more than a fairy tale. "Happy nations, we have all heard, have no history," wrote Louis Kronenberger in the *New York Times*. Stong's Iowa was a mythical land, impervious to political strife and economic downturns. Kronenberger found the novel so unrealistic that it can "teach us nothing significant about life." Even more withering criticism came from Robert Cantwell in the *New Republic*, who observed dryly that "Mr. Stong's dreamy Iowa would seem an even more appealing land if we did not have so much evidence indicating quite clearly that it does not exist."[24]

Critics of the novel were not killjoys. American farmers suffered terribly in the early thirties and sometimes lashed out angrily against whoever and whatever might plausibly be blamed for the Great Depression. Thousands of Iowa farmers had lost their land to foreclosure during the Depression. In 1931 the infamous "Cow War" erupted in southeastern Iowa, home to the Stong and Frake clans, when scores of farmers threatened to resist forcibly the state government's effort to eliminate bovine tuberculosis, fearing that their herds would be declared unfit for market and destroyed. Governor Dan Turner dispatched three regiments of National Guardsmen to the rebellious counties, and the "war" ended without casualties. The following year, the Farmers' Holiday Association urged farmers to "strike" by keeping their crops off the market in order to pressure the federal government to guarantee higher commodity prices. Although the strike foundered amid farmers' chronic inability to organize and engage in collective action, the upsurge of rural discontent convinced some Americans that an insurrection might erupt in—of all places!—the rural Midwest. Despite the ravaged farm economy, Stong never wavered in his insistence that most farmers remained prosperous and content, and he boasted that he had launched a "one-man revolt" against the bleak novels of rural life, stretching from Hamlin Garland's *Main-Traveled Roads* to Erskine Caldwell's *Tobacco Road*.[25]

Although *State Fair* is devoid of specific references to the political and economic problems facing rural Americans in the 1930s, it nonetheless comments obliquely on their plight. Paradoxically, the novel simultaneously discounts the cultural differences that separated rural and urban Americans *and* suggests that farmers and city dwellers remain essentially, immutably different. The Frakes feel no discontent with farm life, harbor no envy toward their neighbors in town, and live on a well-appointed farmstead with modern amenities. In order to emphasize that the Frakes, particularly the younger Frakes, are not hicks, Stong virtually denies the existence of rural-urban cultural tensions. Ultimately, however, *State Fair* suggests that rural and urban life remain utterly distinct. Although Margy and Wayne scarcely seem like farm youth in their dress, speech, or attitudes, both are determined to remain on the Frake farmstead for life. By ignoring the economic problems afflicting rural Americans and insisting that rural folk were happy to remain in the countryside, *State Fair* allayed Americans' worries about the exodus of young people from the countryside, assuring readers that the family farm and the farm family would endure. At the depth of the Great Depression, while Americans hounded another native Iowan out of the White House, Phil Stong offered Americans an archetypal midwestern farm family—cheerful, prosperous, and utterly content with farm life—and the popular response proved overwhelming.

Vaudeville Rusticity

Stong moved to Hollywood to work on the screenplay for *State Fair* and had plenty of news to send the folks back home in 1932: Will Rogers landed the role of Abel Frake, and Fox cast Janet Gaynor as Margy. The supporting cast also included several well-known actors: Lew Ayres as Pat, Sally Eilers as Emily, and Louise Dresser as Melissa. Director Henry King even purchased the state fair's champion Hampshire boar for the role of Blue Boy, and he shot background footage for the movie at the 1932 Iowa State Fair.[26] Stong, however, soon developed an almost allergic reaction to both Hollywood and screenwriting. He considered himself a serious writer, not a studio hack, and he was eager resume work on his second novel, *Stranger's Return*.[27] Hollywood had made him rich—rich enough to spurn Hollywood and return to New York and his writing. As bread lines lengthened and many farmers plunged into bankruptcy, Stong's stock soared: "Isn't the depression awful?" he quipped.[28]

Director Henry King intuitively understood the mythic appeal of Stong's novel. Upon reading *State Fair*, King immediately recognized the iconic allure of the Frakes, the farmstead, and the fairgrounds. Plot and dialogue were almost an afterthought. As King recalled, "The images came first and the sound supplemented them." King rewrote Stong's story in one key respect, adhering to the Hollywood formula by insisting on a happy ending. After their bittersweet parting at the fair, Margy and Pat are ecstatically reunited in the film's final scene. This Hollywood ending appealed to viewers, but it irked Stong and upended the novel's message about the essential difference between rural and urban life.[29]

Critics responded to the film more favorably than to the novel, although some reviewers harrumphed at Wayne's and Margy's sexual encounters and the irrepressibly happy depiction of farm life. Show business magazines predicted that the film would score a hit at the box office. *Billboard* scoffed at the movie's "phony happy ending" but declared that *State Fair* would prove "completely enjoyable to city folk who know nothing of state fairs, and a great deal more than that to country folk who do." *Variety* pronounced the film "a winner all down the line," promising that "those who know their rural America will find it ringingly true." When *State Fair* opened in Iowa in February, the *Register* hailed its "accurate presentation" of the fair, which contained only "a few exaggerations that would be noticed only in Iowa."[30] Some non-

Iowans were less complimentary. Dwight MacDonald dug his finger-nails into the armrests of his seat and resisted the urge to stalk out of theater, but afterward he found it "hard to write with the proper critical restraint." At the depth of a ruinous Depression, he fumed, "'State Fair' is an insulting 'let 'em eat cake' gesture." Stong's "cheerfully trivial sto-ry" looked "about as earthy as the gingham overalls in a musical com-edy number" and made a cruel mockery of farmers as they confronted falling crop prices and foreclosures.[31]

Critics notwithstanding, *State Fair* became a box office hit. The mov-ie received an Academy Award nomination for Best Picture, and the success of *State Fair* and *Cavalcade* (which received the Oscar for Best Picture in 1933) rescued Fox Studio from bankruptcy. *State Fair* also marked the pinnacle of Will Rogers's screen career and was the last hurrah for Janet Gaynor, whose star began to fade afterward. Even Blue Boy became a star, stealing several scenes with his wry, grunted obser-vations. When he died the following year, the celebrity porker was even accorded an obituary in *Time* magazine.[32]

> Died. "Blue Boy," prize hog, film actor, star of the Phil Stong–Will Rog-ers cinema *State Fair*; of overeating and overgrooming; in Hollywood.

The Reluctant Regionalist

Basking in *State Fair's* extraordinary success, Stong had vaulted in only a few months from writing advertising copy to riches and the cusp of a promising literary career. He conceded that *State Fair* was hardly *Main Street* or *The Great Gatsby* but felt confident that his best work lay ahead. He aspired to write about other themes and locales and yearned to take his place alongside Sinclair Lewis and F. Scott Fitzgerald in the pantheon of American novelists. But the dictates of the publish-ing business and Stong's particular talents ultimately prevented him from ranging beyond the bland, purportedly "regionalist" style of his debut novel. Even before *State Fair* rolled off the press, Stong's editor and agent urged him to follow it with "another farm novel." Although he aspired to write about themes other than rural Iowa, within the pub-lishing trade Stong had already been typecast as a local colorist.[33] After he helped launch the vogue for lighthearted tales of rural America, ed-itors urged him to trade on his success by producing more innocuous stories set in the Midwest.

Despite his literary ambitions, Stong showed little aversion to plow-ing the field of rural fiction. His second and third novels, *Storrhaven* and *Village Tale*, reiterated many of the themes and locales of *State Fair*, and mingled autobiographical and historical details.[34] With three novels to his credit, however, Stong yearned to leave rural Iowa behind and write a story set in an entirely different milieu. He wrote to one friend in 1934 that "confidentially, I'm trying to get away from the pig and pickle and playboy stuff and see what I can do on my merits. In other words, I'm trying to write a book that will be sufficiently sound to make people forget that I wrote 'State Fair.' I'm not ashamed of the little story, but I'd hate to stand on it."[35]

In his fourth novel, *Week-End*, Stong attempted to break out of the role of farm novelist by stealing a few pages from Fitzgerald's withering exposé of corruption and high society, *The Great Gatsby*. *Week-End*, takes place in Connecticut, where Stong had recently purchased a country estate. Socialite Flora Baitsell celebrates her thirty-third birthday by inviting a few of her wealthy friends for a get-together at her country home. After two days of drunken revelry, criss-crossing sexual attractions and jealousies, and a fatal automobile accident, the stunned partygoers contemplate life's brevity and resolve not to postpone happiness a moment longer. By *Week-End*'s end, each has paired off with a new lover.[36]

Stong's escape from literary Iowa lasted no longer than *Week-End*. The novel's sordid tale made readers nostalgic for the well-scrubbed wholesomeness of the Frake farmstead. Reviewers panned the novel, and as Stong conceded, "The New York papers all said 'back to the farm.'" He even worried that his attempt to shed the tag of regionalist author would prove a fatal career misstep, from which he might never recover. After this critical rebuff, Stong jettisoned his literary ambitions and reconciled himself to reworking the themes, characters, and locales developed in his first three novels. He reconciled himself to a career as a commercially successful but minor author and became one of the most prolific chroniclers of life in Iowa, churning out a series of novels about midwestern life over a career lasting a quarter century. Ultimately, he produced eight more novels set in southeast Iowa and proclaimed himself a "regionalist," albeit a reluctant one.[37]

Return to the Fair

In the course of his career, Stong repackaged virtually every aspect of his family's and Iowa's history between hard covers in a series of novels and historical works about Iowa.[38] Despite his prolific output, however, only *State Fair* earned renown, and his first novel cast the die for his en-

tire literary career. *State Fair* continued to attract readers and moviegoers for decades. In 1952, twenty years after publishing the novel, Stong
wrote its sequel, *Return in August*. Two decades after Henry King had
irked Stong by adding a Hollywood ending to *State Fair*, Stong wrote
a belated happy ending of his own, allowing Margy and Pat to be reunited after Harry Ware is crushed to death in a tractor accident. *State
Fair* also inspired a Rodgers and Hammerstein film musical, released
by Fox in 1945, and another musical (relocated to the Texas State Fair)
starring Ann-Margret and Pat Boone in 1962. Stage versions of the musical continued to enjoy success on Broadway for decades.[39]

When Stong died suddenly in 1957, the *New York Times* credited him
with struggling "industriously but vainly" for a quarter-century to duplicate *State Fair*'s success, but noted bluntly that "he disappointed
serious critics," who "felt that he had betrayed his talents." *The Des
Moines Register* was kinder, praising Stong as a hometown boy made
good and one of Iowa's best-known native sons. Like Abel Frake, Phil
Stong attained his cherished goal, but "They" had gotten the best of
him. After *State Fair*'s phenomenal success, Stong seemed poised to realize his dream of becoming a significant writer, but soon found himself fenced in by the vogue for rural, ostensibly "regionalist" fiction that
he had helped to create. Genuinely fond of Iowa, he banged out novel
after novel set in Iowa because neither his talents nor the dictates of the
publishing business would permit him to do otherwise. Despite Stong's
valiant effort to escape, his publishers and readers effectively kept him
down on the farm. In the thousands of pages he published during his
prolific career, Stong scrubbed the rural Midwest of any idiosyncrasies
or troubles that might render it unrecognizable or unappealing to a
mass readership and assured his readers that the family farm would
endure unchanged in an urbanized and industrialized America. In his
determination to craft an appealing image of the Midwest, Stong rendered the region, farm life, and small-town society bland and neatly
packaged them to suit the tastes of Americans everywhere.[40]

The Fair on Canvas

At the same moment that *State Fair* offered Americans a cheerful
page-turner about rural Americana, Grant Wood and a group of avowed-
ly "regionalist" painters created their own images of the Midwest and
staked an ambitious claim for an indigenous regional culture. The state
fair's Art Salon, by far the state's most closely watched art gallery, be-

came the main venue in which these painters displayed their work and launched the vogue for regionalist painting. Regionalist paintings immediately caused a stir not only at the fair, in Iowa, and in the Midwest, but across the nation. Did regionalism mark the arrival of a distinctly American style of painting? Or was it a reactionary aesthetic that flew in the face of modernist art? Did regionalists offer an appealing image of life in the Midwest, or were there canvases unflattering, even condescending? During the 1930s, debates over paintings provoked impassioned debate over the role of artistic culture in the Midwest.

Judging art proved even trickier and more controversial than judging livestock, crops, machines, foods, or handicrafts. Judges decided which paintings would win ribbons, but exhibitors and fairgoers also felt entitled to form their own opinions. As a writer for *Wallaces' Farmer* remarked after the 1930 fair, visitors to the art exhibit frequently discovered, sometimes to their dismay, that the art judge did not share their taste in paintings. Judges might be knowledgeable about painting and drawing, yet art could not be judged according to the ostensibly scientific criteria employed in the agricultural contests. The art contest remained inescapably subjective: the judge's taste invariably played a huge role in determining the winner. Art judging encompassed both subjective opinion and established canons of aesthetics, and art could be either a repository of traditional values or an unsettling harbinger of cultural change. As a result, fairgoers considered the fair's art exhibit an especially sensitive barometer of Iowans' cultural attainments and values. At stake in the art competition was much more than ribbons, and the display of paintings ranked among the fair's most closely watched exhibits.

The Crowning Glory of the Fabric

The fair's art contest offered an especially telling glimpse of both the progress and difficulty in creating a vibrant culture, built atop agriculture, in the Midwest. Fine arts occupied a unique position among the fair's myriad exhibits of agricultural products and useful "arts." The founders of the state agricultural society earnestly believed that their efforts to improve the state's agricultural economy would lay the groundwork for the eventual development of other industries and the arts, and so they included prizes for painting in the premium list for the first state fair in 1854. Miss Jane Funk of Jefferson won the art competition at the inaugural state fair, receiving a premium of one dollar for exhibiting the "best floral painting." Painting, in the fair's classification of premiums, was but one among many "arts," and was not completely differentiated from skills and crafts. Relegated to the final category (variously termed "Inventions, &c." or "Miscellaneous") in the fair's premium list, the competition for the "best oil painting" was typically lumped together with "best improvement for roofing houses" and "best lot of pressed brick." Listing paintings in the final category of the premium list, however, also indicated that painting and other arts were regarded as the very pinnacle of the fair's exhibits, the tiny but important apex of a cultural pyramid built atop a massive base of livestock, crops, and machinery.[41]

On a fairgrounds filled with displays of livestock and farm implements, the art exhibit offered one of the few exhibits of particular interest to women, and the majority of contestants were women, especially in the fair's early decades. Painting, along with other arts and crafts, was considered a refined, leisurely pastime, reserved for those who did not have to expend their energy in agricultural production. Painting, unlike growing crops, was not utterly necessary to survival, but nonetheless merited esteem as an elevated and refined activity, one that attested that Iowans could do more than raise corn and hogs, and could build a culture atop agriculture.[42]

Fairgoers certainly did not neglect the arts but crowded into the Fine Art Hall. Most Iowans could see cattle and hogs any day of the year, while they seldom enjoyed the opportunity to view paintings. But the art exhibit's popularity cannot be attributed to mere novelty. Iowans commonly regarded artistic attainments a measure of the state's cultural progress, and they scrutinized the art exhibit closely in order to gauge their state's development. After viewing the 1859 art exhibit, a reporter for the *Northwest Farmer* wrote confidently that "we shall expect to see these and other branches of the fine arts flourish in our State, in proportion to the increase of settlements, wealth and refinement." In the words of another agricultural journal, the exhibit displayed nothing less than a compendium of "the all of everything that our State has attained in the beautiful, ingenious or intelligent." Observing with satisfaction the enormous interest in the arts displayed at the 1882 fair, the *Iowa State Register* contended that the art exhibit offered "the attraction par excellence, the crowning glory of the fabric," and the sole exhibit in which all Iowans could find common ground among the fair's diverse activities.[43] Farmers thronged the crop and livestock exhibits, mechanics scrutinized machinery and inventions, and housewives marveled at displays of cookery and handicrafts. Their paths crossed in the fine art exhibit, which satisfied a shared "craving" for beauty "which neither the sight of the production of the soil, the herds of animals, the half-living machinery, the beautiful flowers, nor the speed of the fastest animals, can ever supply."[44]

The experience of viewing art at early fairs underscored the comparatively undeveloped status of painting in Iowa. A visitor to the Fine Art Gallery at the 1856 fair, for example, entered through the door on the north, passing by displays of embroidery, needlework, quilting, reptiles and insects, a two-headed calf, photographs, penmanship, stone carving, surveyors' instruments, "various school apparatus," including an orrery, a tellurian, globes, and a gyroscope, before at last reaching the paintings. The organizers of the fine art exhibit may have intended to suggest some sort of cultural hierarchy through their arrangement of the exhibit; on the other hand, it is also possible that the paintings came last simply because they were hung on the wall behind the tables laden with other items. The art exhibit could sometimes be even less well organized. The art judging committee at the 1867 fair filed a complaint afterward, urging "that in the future all articles of the same class be grouped [together]. It is a serious inconvenience, and one leading to very uncertain results, to have one article in any given class at one end of the hall and another competing with it placed at the opposite end." Problems with the art exhibit persisted for decades: in 1889, art judge N. B. Collins bemoaned "the great difficulty attending to the work of judging the art display of the fair . . . with the competing pictures hung on walls 15 to 20 feet high and scattered and mixed like needles in a haystack."[45]

The qualifications of the judges themselves were another frequent source of dissatisfaction with the art exhibit in the fair's early decades. Many of the art judges, like those in other divisions of the fair, lacked any particular expertise their field and were often selected simply because they consented to perform a difficult and thankless task. Judges often had to be rounded up on the fairgrounds, and sometimes at the last minute. The fair hired professional judges to oversee its livestock competitions in the 1870s and 1880s, but the art exhibit continued to be judged by amateurs into the early twentieth century.[46] Numerous other problems confounded the agricultural society's effort to mount a satisfactory art exhibit. The grandly named Fine Art Hall was invariably a hastily constructed, floorless, wood-frame building that exposed paintings to weather, dust, and the threat of fire. Agriculturists, journalists, and fairgoers alike frequently asserted that the art exhibit would remain disappointing until a suitable building was created to house it.[47]

For all of these reasons, critics grumbled for decades that the art show's entries were spotty, that judges sometimes lacked clear criteria for awarding premiums, and, worst of all, that the art exhibit lagged far behind the fair's other displays. To the distress of many observers, artistic progress failed to keep pace with the state's rapid material growth. In 1873 the *Western Farm Journal* stated that the Fine Art Hall "justly demands most attention from visitors because most defective in comparison to its importance to our future development." The journal complained that Iowa's artists remained unsuccessful because they attempted to copy schools of painting from New England and Europe, instead of utilizing the same source of strength that produced the state's agricultural bounty. As a result, the paintings on display were "insignificant, totally unworthy of comparison with the results of other branches of Iowa education. . . . Almost all of them inevitably copies from others and inevitably inferior to them. Feeble attempts to re-produce the fashions of feeble art schools, but not a single effort at original design or execution." In art as in agriculture and manufacturing, many midwesterners felt, the state ought to strive for self-sufficiency, rather than remain beholden to styles that originated in distant cities.[48]

"With the exception of paintings," noted the *Iowa State Register* in 1879, the fair's "display is equal in every respect to those of other exhibitions, and superior in many respects." Three years later, the *Iowa Homestead*, after surveying the extensive array of items in the Fine Art Hall, wondered, "But where was the fine art? Certainly not in the paintings, with the exception of a few of the water colors, and *possibly* half a dozen of the oil paintings [emphasis in original]."[49] Despite sporadic attempts to improve the quality of the art show, fairgoers, fair officials, and journalists alike generally regarded it the fair's weakest department for decades.

In the 1890s, efforts to improve the art exhibit gained momentum, as a growing number of Iowans became convinced that the state deserved a fine art exhibit worthy of the name. In 1889 fair officials divided the exhibit into professional and amateur classes, distinguishing trained artists from hobbyists. The following year, according to the *Register*, brought a number of improvements, "not only in the character of the work, but especially in the management." For the first time in the fair's history, the exhibit had been carefully classified and arranged, "making it possible for the person awarding the premiums to see the work side by side and thus be enabled to judge more correctly of the comparative merits of each."[50] These improvements inspired hope that Iowans would begin to demonstrate progress in the arts, as they had in agriculture. The *Register* remarked in 1891 that most of the art exhibit was quite good, "while only a few years ago most of it was positively bad." The following year, the newspaper noted, reassuringly, "that there are no atrocities this year exhibited, such as portraits painted in frying pans or landscapes in bread bowls and on shovels" or other examples of folk decoration to detract from the framed canvases of real art.[51]

Despite glimmers of progress, however, complaints about the quality of the art show and the competence of its judges continued to vex the fair for years. These complaints grew in number and influence as more Iowans became impatient with the state's sluggish cultural progress.[52] An editorial in the *Des Moines Leader* in 1896 brooded that

> the display of drawings and paintings at the state fair this year was of such a character as to set one thinking of the condition of the pictorial art in Iowa. In no other department of the fair could there be found such a medley of good and bad. Some of it was in touch with the most advanced ideas of the times and some with the extreme backwoods of a primitive civilization. It is to be regretted that the management does not discriminate and encourage only that which is in direct line with the art world; that which could be recognized by the institutions which lay the foundation for real art and pass judgment upon the finished product.[53]

The editorial accused state and county fairs of committing "crimes against the progress of art" and called for a thorough examination of the method of selecting and judging entries to the fair's art show, so that the exhibit could adhere to the same standards as established art museums, "the institutions which lay the foundation for real art."[54]

In the editorialist's view, only one Iowan possessed the credentials to uplift the condition of art in the state: Charles Atherton Cumming. An accomplished academic painter, Cumming had studied at the Chicago Academy of Design (later the Art Institute of Chicago) and the Academie Julien in Paris. Upon completing his training, Cumming returned to Iowa, where he founded the Department of Fine Arts at his alma mater, Cornell College, and later the Department of Art at the University of

Iowa. His abiding concern, however, was his own private school of art in Des Moines, where he taught academic drawing and painting to legions of devoted students.[55]

Cumming had firm opinions about painting and the state fair's role in fostering art in Iowa, which he outlined in a letter to the *Leader* in 1897. "We do not want a standard set before us that is a relic of some old time country fair," he wrote, arguing that "out here in Iowa the only reasonable thing for us to do is to follow the lead set by those who know." For Cumming, this meant emulating the work of French salon painters. Iowans might participate in "high" culture, but they were in no position to originate it. Cumming recognized that the state fair's art competition, for better or worse, was by far the state's most influential art exhibit, and he felt that, with a few improvements to its rules and judging, the fair could advance Iowa's artistic development. He insisted that an attractive and permanent art building be constructed in order to house the exhibit. Even good paintings, he said, looked "miserable" in the dimly lighted, barnlike art hall. Cumming also insisted that the fair abolish the distinction between amateur and professional painter, so that all entrants would be judged according to a single standard of excellence. If these reforms were implemented, Cumming predicted, "the best class of painters in the state would be seen at the fair."[56]

In 1899 the Agricultural Society decided, in the words of the *Leader*, "to have a real art show or else give up the effort to have any at all," and commissioned Cumming to overhaul the fair's fine arts department. He responded by devising new rules for the exhibit, and he persuaded the society to erect a hall for the artworks, so that they could be displayed separate from other exhibits. The comparatively unimposing Fine Art Hall was built adjacent to the fair's much larger Exhibition Building, atop the fairgrounds' hill—yet another reminder of art's ambiguous status as both the pinnacle and one of the least developed of the fair's exhibits. Cumming promised hopefully in February 1899 that, as a result of his reforms, "the horrible nightmares that have offended the artistic eye of the state under the guise of 'art' in the art hall probably will be seen no more."[57]

Despite Cumming's efforts, however, the art show continued to disappoint many fairgoers. Fair Board secretary Arthur Corey remarked bluntly in 1913 that "it is a well-known fact that we have been unable to induce anyone to exhibit anything of value or worthy of mention," recommending that the art exhibit once again be moved from its wood-frame building (commonly derided as the "Art Barn") to the new Women's and Children's Building, along with yet another "thorough revision of the classification" of the art show.[58] Corey appointed Cumming superintendent of the art exhibit, and Cumming immediately set out once more to improve the display. When the fair had ended, the *Register and Leader* proclaimed that the

ART EXHIBIT AT THE FAIR SETS
NEW STANDARD FOR THE STATE[59]

Under Cumming's tutelage, paintings would no longer hang in an "Art Barn" or in galleries stocked with displays of dental fixtures and stuffed sparrows but would grace the walls of the fair's Art Salon. As the catalog for the 1916 fair put it, "Of late years the rules and conditions surrounding the exhibit have been made a little stricter, the lines gradually drawn a bit closer, the idea being to exclude all productions unworthy of merit and to raise the standard of the exhibit."[60]

Cumming sought to use the fair's art exhibit to ensure that the aesthetic standards of academic painting would govern Iowa's most influential art gallery. In the wake of modern art's controversial debut in America at the 1913 Armory Show in New York, ripples could be felt as far away as Iowa, and Cumming fiercely defended academic painting in the United States and sought to insulate Iowans from "the passing freakish tendencies in 'modern art.'" He believed fervently that paintings ought to be representational depictions of conventional subjects and that artists should adhere to established aesthetic canons governing perspective, line, and color. Although he subscribed to the aesthetics of European academic painting, Cumming also believed that Iowa had advanced far enough culturally "to justify its people in making its own expression and recording its own history." The best way—indeed, the only way—for Iowans to create their own art was to emulate European salon painters and reject impressionism, cubism, and all styles of modernist painting.[61]

Cumming presided over the state fair's Art Salon from 1913 to 1926, during which time he became an increasingly virulent opponent of all styles of modern art. Iowans had struggled for decades to build their economy and elevate their culture, and Cumming frequently likened fine arts to the apex of a pyramid based on a broad economic foundation. Just as Iowans' were set to place the crown atop that cultural pyramid, according to Cumming, proponents of modern art threatened to subvert Western culture altogether. He characterized the effort to build civilization and culture in bluntly racist terms, calling academic paint-

ing the "white man's art" and insisting that it represented the highest
attainment of human civilization. He credited academic painting, with
its well-defined standards of line, light, color, perspective, and subject,
with liberating the white race in Europe and the United States from
the inferior "symbolic" art created by "primitive races" and utilized
in "modern" painting. Cumming fretted that most Americans were
too ignorant about art "to defend themselves against this encroaching
blight" of modern art, and he accused modern artists of infiltrating the
state's universities and even attempting to commandeer his own art
school in Des Moines.[62]

The state fair's Art Salon soon became a battleground in the war
between traditional and modern artists. In 1925 the growing rift between academic and modern painters in Iowa was exposed when Cumming abruptly postponed the fair's art judging, because, he claimed,
the judge, Frances Cranmer Greenman of Minneapolis, had suddenly
fallen ill. When Greenman personally disputed Cumming's diagnosis,
Cumming disclosed the true nature and gravity of her affliction: "Mrs.
Greenman has been converted to what she calls 'modern' art since I
last viewed her exhibit. Here in Iowa our artists are followers of 'white
man's art,' which is directly opposed to the modern 'jazz' art."[63]

Greenman's replacement, J. Laurie Wallace of Omaha, received a
clean bill of health from Dr. Cumming. "I have no sympathy for most
of this modernist stuff," Wallace declared, predicting that artistic
"vogues" such as Picasso, Matisse, and Gauguin would soon be deserv-
edly forgotten, as artists began once again to paint canvases "conform-
ing to nature, representative, and holding within their own field." To
academic painters such as Wallace and Cumming, cubism and expressionism not only violated established artistic canons but threatened to
undermine art's worth as a repository of timeless aesthetic values and
as an index of the level of civilization. In modern art, Wallace declared,
"standards are demoralized, and there is no criterion for evaluation.
Lines of demarcation break down. It is argued that you can't judge such
works on the basis of color, form, line; so what sort of artistic standards
must one fall back on?" Without well-established criteria as to what
constituted a beautiful or accomplished painting, art could no longer
embody excellence and civilization, and beauty would truly be in the
eye of the beholder.[64]

After the 1926 fair, Cumming retired to California, where devoted
himself to his writings on art, convinced that he was "inspired as a

prophet with a message for the world." He outlined, but did not complete, two screeds, *A Defense of the White Man's Art* and *Democracy and the White Man's Art*. After Cumming's departure, conservative painters briefly maintained the upper hand at the fair, and the Fair Board entrusted the Art Salon to another conservative artist, Zenobia B. Ness of Ames, who taught drawing at Iowa State College. During Ness's first two years as head of the Art Salon, Cumming's students continued to dominate the competition as they had for years. Then, in 1929, "modern" art took over the salon, arriving not from Paris or New York but from Cedar Rapids.[65]

From Paris to Stone City

In 1928 a distinguished elderly resident of Cedar Rapids sat for his portrait. His portraitist, local artist Grant Wood, depicted his subject standing before a frontier era map of Linn County. Wood's *Portrait of John B. Turner, Pioneer* suggested the enormous changes that had transformed Iowa within Turner's lifetime. The resolute expression on Turner's face and the familiar grid of the midwestern landscape printed on the map both attest to the fortitude of the state's pioneers and their success in transplanting civilization to the frontier region. Wood's portrait of Turner heralded the advent of a new midwestern school of painting that would soon be dubbed "regionalism." Eager to promote his work, Wood entered the painting in the only forum where it would be viewed by thousands of Iowans, the Art Salon at the 1929 state fair. It was awarded the fair's grand prize, launching not only Wood's career but a new school of American painting.[66]

As an aspiring young artist, Wood had admired French impressionism and had spent several years in the 1920s studying painting in France. During a trip to Belgium and Germany in 1928, he was smitten by the austerity of Flemish painting, and upon his return to Iowa he began to depict midwestern subjects and landscapes in a similarly spare style, while simultaneously proclaiming his determination to break free from the dominance of European art and create an authentically American art. American artists, Wood declared, should stop trying to paint Parisian streetscapes or the Provençal countryside, but should instead depict familiar subjects from their own locale. Regionalism, as this indigenous school of art came to be called, was most famously associated with Wood, Missourian Thomas Hart Benton, and Kansan John Steuart Curry, but it included scores of largely unknown artists as well. Its rapid growth and influence suggested that Iowans had at last created a home-grown culture, one deeply embedded in the state's heritage and its agricultural economy.[67]

As the only art gallery in which most Iowans ever set foot, the state fair's Art Salon offered an ideal venue for Wood's scenes of midwestern life and his effort to gain support for his regionalist aesthetic. Wood's canvases would have earned him expulsion from the Cumming School of Art but won immediate acclaim in Iowa and dominated the Art Salon's competitions in the early 1930s. In 1930 he won both the grand prize and first place in the landscape category for *Portrait of Arnold Pyle* and *Stone City*, his landscape of the future site of his regionalist artists' colony. In that year, both *Stone City* and *American Gothic* were selected for the Exhibition of American Paintings at the Art Institute of Chicago. *American Gothic* became an almost immediate sensation, and Wood became a renowned artist. As he wrote to Zenobia Ness, superintendent of the fair's art exhibit, credit for his sudden success lay "with you and the state fair." The following year he won again with *The Appraisal*, a comic juxtaposition of a plump, stodgy city woman and a radiant, happy farm woman. In 1932 Wood extended his string of grand prizes with *Fall Plowing*, a nostalgic landscape foregrounded by a horse-drawn plow, an implement now consigned to obsolescence by the tractor. After four consecutive victories, Wood retired from the fair's art competition, although he continued to exhibit works at the fair through 1934.[68]

As the leader of a growing regionalist movement in Iowa, Wood attempted to consolidate the prestige and influence of his new school of painting. In 1932 Wood, Adrian Dornbusch, and Edward Rowan established their renowned artists' colony at Stone City, northeast of Cedar Rapids. For two summers this tiny town, its physical setting dominated by an enormous quarry, became the epicenter of Iowa's artistic scene, as Wood and a remarkable collection of faculty and students sought to create a haven in which regionalism could flourish. After only two years, however, Wood left the colony to assume the directorship of the New Deal's Public Works of Art Project in Iowa and accept a professorship in the Department of Fine Arts at the University of Iowa.[69]

Despite Wood's success and popularity—he had by this time become famous nationwide and had assumed Cumming's mantle as the state's most influential artist—regionalism provoked considerable controversy both in Iowa and throughout the United States. Although Wood denied accusations that he sought to impose his own style on other midwestern painters, he did strive to promote his distinct vision of regionalism. In 1935 he published a regionalist manifesto, *Revolt Against the City*, ghostwritten by Frank Luther Mott, a professor of literature at the University of Iowa. The pamphlet asserted that, at root, regionalism entailed "an honest reliance by the artist upon subject matter which he can best interpret because he knows it best." As the pamphlet's title declared, however, regionalism was not merely a new artistic style but a rejoinder to urbanization generally. Like the group Twelve Southerners, whose regionalist manifesto, *I'll Take My Stand* (1930), inveighed against a homogenized American culture, Wood considered regionalism as a gesture of defiance against urbanization and a national, mass culture.[70]

Wood's call to "revolt" against cosmopolitan art and urban culture, like many manifestos, overstated its case for rebellion and reduced its enemies to straw men. Wood's paintings imported their spare style and use of light from Flemish painting and borrowed some of their highly stylized, almost geometric, composition from modern design. Despite Wood's insistence on painting local subjects, he shared with most regionalist painters a keen interest in artistic currents in the East and in Europe. Midwestern regionalism never became an entirely coherent artistic style or movement. Thomas Hart Benton used bold color, sinuous lines, and a heady sense of movement and space to evoke an ambivalent attitude toward the Midwest, while John Steuart Curry developed a brushy style to paint vignettes set in his native Kansas. Still, Wood's manifesto expressed an important truth about regionalism: the new school of painting defined itself primarily by what it was *not*. Regionalism could not be codified into a single stylistic, thematic, or political program. Instead, it represented the declaration of cultural independence by midwestern painters, who defied the established arbiters of the art world and insisted that they could find as much inspiration in an Iowa cornfield as Monet found in the gardens at Giverny.[71]

Regionalism swept the art world like a cyclone in the 1930s, provoking impassioned debate in both art journals and popular magazines. Thomas Craven, art critic for the *New York American*, became regionalism's most ardent champion, hailing the new style as a tonic for Americans' cultural inferiority complex and as art suited to the rough-and-tumble American environment. But detractors were numerous and influential, and regionalists immediately found themselves assailed by artistic conservatives, modernists, and leftists. In conservatives' eyes regionalist paintings were at best amateurish, or even downright subversive of artistic standards. Modernists considered regionalists' representations of the countryside antithetical to a cosmopolitan aesthetic, while left-leaning critics derided regionalists' sentimental, bucolic vision of farm life for obscuring the hardships endured by rural Americans during the Great Depression. Stuart Davis, an accomplished modernist painter and art critic, dismissed regionalism as little more than a "slight burp" in American art, but one symptomatic of "the stomach ulcer of Fascism."[72]

Regionalism stirred controversy in Iowa as well, especially at the state fair. In 1933 art judge Rene d'Harnoncourt awarded blue ribbons to Wood's Stone City colleagues Adrian Dornbusch and Arnold Pyle, in the oil and watercolor competitions. Harnoncourt hailed their works as proof that an indigenous American art had arisen in the Midwest, praising them as images "directly reflective of the locality."[73] The following year, however, regionalists suffered an apparent setback when superintendent Zenobia Ness appointed Louis Le Beaume, president of the St. Louis Art Museum, to judge the art contest. Many regionalists considered Le Beaume's tastes too conservative and protested his appointment as art judge. Rapidly growing in number and influence, regionalists insisted that the fair ought to select judges who favored the new school of midwestern painting. Le Beaume ultimately surprised his critics by awarding the grand prize to a regionalist work, *The Butchering*, by Thomas Savage of Fort Dodge, a farmer who had studied painting at Stone City.[74]

In an attempt to mollify regionalists, Ness decreed that henceforth the artists themselves would be allowed to elect the judge of the Art Salon, a policy that, whatever its merits, could only make the subjective and factionalized practice of art judging more apparent to fairgoers. Ness's decision, ostensibly an effort to appease regionalists, was in fact a shrewd piece of electioneering, because it contained a "grandfather clause": any artist who had ever entered the fair's art show was allowed to cast a ballot for the art judge. Although the number of regionalists had grown rapidly in the 1930s, they were far outnumbered by the conservative artists who had predominated at the fair for decades. When Ness tallied the ballots in 1935, Frederic Tellander, a conservative Chicago artist, had defeated regionalist candidate Dewey Albinson of Minneapolis. As *The Des Moines Register* reported, the outcome reflected a party-line split between alumni of the Cumming School of Art and regionalists. To their dismay, however, conservative artists soon discovered that Tellander did not subscribe to the artistic spoils system when he awarded the grand prize to a regionalist work, *River Bend*, by Marvin

Cone of Cedar Rapids, another of Wood's colleagues from Stone City, and second prize to *Country Gas Station*, by regionalist Harry D. Jones of Des Moines. Conservatives had elected Tellander as one of their own, only to discover that he was a maverick.[75] Alice McKee Cumming, widow of the conservatives' mentor and herself an accomplished painter, complained bitterly during the 1935 fair that the growing number of regionalist paintings on display demonstrated that Iowa's artistic development had actually slid backward in recent years, degenerating into "a homespun county fair art show, glorified by the name the Iowa art salon." The state's "real painters," she claimed, no longer competed at the state fair but "prefer to present their works in dignified exhibitions in cultural centers of the East."[76]

In 1935, in addition to allowing painters to select the art judge, the Art Salon also polled fairgoers' artistic preferences, and the results proved reassuring to traditional painters. In yet another attempt to circumvent the judges' preference in recent years for regionalist works that many Iowans did not appreciate, Ness created a "popularity prize," to be awarded to the painting receiving the most votes from the salon's visitors. Some 80,000 people, roughly one-quarter of the fair's visitors, cast ballots. When they were counted, according to *The Des Moines Register*, Iowans had "evened the score" with Judge Tellander, choosing quaint paintings of spinning wheels and swans over Cone's and Jones's works. "Angular paintings by modern artists may win prizes from the Iowa state fair art salon's official judge," the newspaper declared, "but results of a popularity ballot proved the tall corn state's citizens still like their pictures lifelike, pretty and conservative." Although academic painters had never staked the value of their work on popular approval, they were gratified to learn that the regionalist aesthetic was not popular with all Iowans. The discrepancy between the tastes of the art judges and the majority of Iowans persisted throughout the decade, but the popularity prize did not. The popularity prize, at odds with the fair's longstanding practice of entrusting expert judges to discern excellence and award premiums, was discontinued after only one year.[77]

Political controversy continued to swirl around the Art Salon in 1936, when the Iowa Co-operative Artists, a regionalist artists' union led by Francis Robert White of Cedar Rapids, boycotted the exhibit, claiming that fine arts should not be judged, which they decried as "gambling" for prizes. The union insisted that artists deserved to be paid a fee for exhibiting their works. The Co-operative Artists' paintings shared more, both aesthetically and politically, with Social Realism than with the archetypal midwestern scenes and light satire favored by Grant Wood and most regionalists. Wood's influence continued to prevail at the fair's Art Salon, however, as his protégé, Arnold Pyle, won both the grand prize for his painting *Big Hooks* and first prize in the landscape category.[78] Dewey Albinson, the "modernist" candidate defeated in the 1935 election, won election as art judge, and awarded Francis White (who abandoned his boycott of the Art Salon) the grand prize for *Ages of Man*, described by the *Register* as a "starkly realistic street scene of poorly clad characters, workmen, and impoverished women—the sort who pass by the lower priced markets of a city." According to *Wallaces' Farmer*, the painting "created much comment, some finding the picture wholly grotesque, others recognizing the figures as types seen on farms as well as on city streets."[79] Regionalist painting had gained a measure of acceptance unimaginable only a few years earlier, but many Iowans still regarded regionalist landscapes oddly unrealistic and their depictions of farmers and working people downright unflattering.

The Founders of Civilization?

In 1938, the controversy over regionalism became even more acrimonious when Dan Rhodes of Fort Dodge won the grand prize for *Painters*. Rhodes's canvas, which he later characterized frankly as "proletarian," depicted not artists but two workmen carrying a ladder.[80] *Wallaces' Farmer*, displaying its usual bemusement toward regionalism, commented that "there were a lot of low-brows who honestly wondered how long it takes for culture to work, as they looked at . . . 'Painters.'"[81]

Rhodes unveiled a much larger work at the 1938 fair, and it generated a correspondingly larger controversy. Along with Howard Johnson, he was selected to paint a 218-foot mural of Iowa's history in the fair's Agricultural Building. Commissioned by the federal government's Works Progress Administration (WPA) to commemorate the centennial of Iowa's accession as a federal territory, the mural depicted the familiar story of white Americans settling the frontier and transforming the wilderness into civilization, an enterprise that the fair had promoted and extolled for decades. The mural commences with pioneers driving away the Indians, surveying their newly conquered land, plowing their fields, and building their homes. Near the center, a farmer sows grain with his left hand, a seemingly innocuous gesture that provoked many Iowans to complain to the Fair Board that Rhodes included this sinister figure as a leftist political symbol. But Iowans were even more concerned that the mural offered an unflattering depiction of the state's history, and that it depicted Iowans as downtrodden, not left-handed. According to *Wallaces' Farmer*, "Farm people studied the pictures diffidently. It seemed queer to see farmers made the heroes of enormous drawings, to see farm men and women—some of them pretty hard-looking, too—done on a scale and with colors hitherto reserved for ladies swathed in cheesecloth, representing the Spirit of Liberty."[82] Fair officials demanded that the WPA order Rhodes to "correct" the picture, but the Iowa division of the WPA's Federal Art Project was now under the direction of regionalist Francis Robert White, who refused to infringe upon Rhodes's artistic freedom.[83]

Unable to compel Rhodes to alter the mural, the Fair Board made an alteration of its own shortly before the 1939 fair opened: it had the painting captioned with Daniel Webster's often-quoted aphorism, "When tillage begins, other arts follow. The farmers therefore are the founders of civilization." Webster's words had been invoked regularly at the fair since George Dixon's inaugural address in 1854, and they might justifiably be regarded the fair's unofficial credo. The addition of Webster's remark served to tame Rhodes's depiction of Iowa's history by framing it as a paean to the state's material progress and to the farmers who lay at the root of that progress. One regionalist, incensed by the alteration of Rhodes's mural, responded angrily that "in the first place, the quotation isn't true. Cavemen drawings, representing a highly developed type of art, have been found in Spain. All evidence indicates these primitive people didn't know anything about soil tillage, and yet they were talented artists." This statement was tantamount to heresy, the very antithesis of the agrarian myth and the mission of state and county fairs. Was agriculture *not* the foundation of civilization? Few Iowans would have accepted such a claim. Regionalists had frequently proclaimed their desire to create an indigenous art, one that would spring from rural life and the land itself and garner the approval of midwesterners, but artists and the public remained bitterly divided over regionalist depictions of midwestern history and life.[84]

By the late 1930s, regionalist painters, however controversial, were securely in control of the fair's Art Salon. "Almost everybody's a regionalist now," declared Rhodes in 1939. Indeed, the definition of "regionalism" had become broad enough to encompass almost everybody, ranging from artists who painted iconic visions of rural life to those whose images of workers and city life could just as accurately be called Social Realism. Rhodes continued to dominate the art show in 1939, when art judge John Steuart Curry awarded him the grand prize for *Bulletin*, an ominous depiction of three men poring over the latest newspaper account of the war in Europe. He won again in 1940 for *Hod Carrier*, an austere portrait of a mason's assistant at work.[85]

Regionalist painters scored a final triumph in 1941, when Nellie Gebers won the fair's grand prize for *Prairie Harvest*, but the 1941 contest marked regionalism's last victory at the Iowa State Fair Art Salon. The outbreak of World War II led to the fair's cancellation from 1942 to 1945, but the war's impact on regionalism proved more enduring. War with Germany accomplished what conservative and radical critics could not, provoking a backlash against "provincial" artists, whose works were likened to Nazi propaganda glorifying the Folk and the Fatherland. Stuart Davis's strident remark that regionalism represented the American counterpart to fascist art now gained greater currency. Iowa's preeminent regionalist, Grant Wood, died in 1942, and regionalism, barely a decade old, succumbed shortly thereafter.[86]

Before the fair reopened its gates in 1946, Fair Board secretary Lloyd Cunningham made yet another alteration to Rhodes's and Johnson's mural: he ordered it taken down and sawed into scrap lumber to build shelving and exhibition booths for the upcoming fair. When questioned about destroying an enormous and expensive public work of art, Cunningham snorted that "the mural isn't art, it was WPA. It is an insult to Iowa farmers because it depicted them as club-footed, coconut-headed, barrel-necked and low-browed. . . . It was a joke to have that thing on a fairgrounds that's devoted to glorifying the Iowa farmer and his accomplishments." Cunningham added that he hoped that the Fair Board's decision to scrap Rhodes's and Johnson's "monstrosity" would inspire Iowans to destroy other "so-called art-pieces which were foisted on them" by WPA artists in libraries, post offices, and other public buildings.[87]

Grant Wood's goal of creating and sustaining an indigenous artistic style in the Midwest proved short lived. Regionalists were beset by critics from several directions, confronting an entrenched circle of academic painters, a Fair Board ill disposed toward controversial works of art, dissension within their own ranks, and a public that did not always appreciate purportedly "regionalist" paintings. Although Wood and other regionalists hoped that midwesterners' allegiance would secure regionalism a permanent niche in American painting, the new school enjoyed only a tenuous, brief existence in the volatile, politicized artistic milieu of the 1930s.

Ironically, academic and regionalist painters shared some common ground: both schools of painting earnestly desired to see the fine arts flourish in the Midwest, and both were reactions against rapid, bewildering cultural change. For the academics, this reaction entailed venerating traditional European painting, with its emphasis on technical proficiency, line, color, and light, and its insistence on representational art and conventional subjects. They hoped to transplant this strand of European high culture in the Midwest and to create a hedge against "modern" art of any stripe, including regionalism.

Regionalists, on the other hand, claimed to cast off all things European in their quest to develop a distinctly American art. But the regionalists' disavowal of the tradition of Western painting and their exclusive focus on indigenous subjects marked a reaction against the rapid extension of mass culture into the Midwest. Even those painters whose works departed from Grant Wood's conception of regionalism proudly continued to call themselves regionalists, and to insist that they could find ample artistic inspiration in their own locale. Regionalism represented a powerful, if unsuccessful, assertion of the primacy of place, in this case a rural, agricultural place, as the basis of culture. Building culture atop agriculture had also been the mission of state and county fairs, whose importance as cultural and educational institutions had slipped in the 1920s and 1930s, as the rapid extension of mass media and mass entertainment offered small-town and rural Americans new forms of information and diversion. The state fair's long-running tension between agriculture and entertainment, the extraordinary success of *State Fair*, and the short-lived regionalist experiment were all expressions of Iowans' ambivalent response to the advent of a national mass culture.

onclusion

....

THE FAIR AND IOWA'S HISTORY

Iowans look forward to the state fair and look backward to its history and tradition. The state fair remains among the most significant and eagerly anticipated events on many Iowans' calendars and the state's most renowned institution. The fair itself gestures both toward the future and toward the past. Passing through the fair's gates, visitors have always marveled at the exhibition, which simultaneously showcases innovation and new technologies, yet hearkens back to the state's founding and its history. For more than a century and a half, the fair has offered Iowans an annual display of their attainments over the past year and a measure of their society's development since the frontier era. From the mid-nineteenth century until World War II, Iowans continued to look to the fair as an important agency for promoting agricultural and economic development and as a sensitive barometer of the state's progress. Because they looked to the fair as a microcosm of their state, they took it seriously and debated, sometimes heatedly, its proper role.

Over the course of its history, however, the fair has become less devoted to accelerating the pace of innovation and more deeply rooted in celebrating tradition and continuity. While the fair still looks to the future and fosters progress, its basic elements remain largely unchanged after a century and a half of almost unimaginable economic and social transformation from the 1850s to the beginning of the twenty-first century. The fair has now become a familiar, reassuring annual ritual, a touchstone of stability amid a world of ever-quickening change. The state fair's history, then, recounts a long-running conversation about vast cultural and economic changes, about tensions between rural and urban America, about agricultural production and commercial entertainment, and about the ties that link past, present, and future. In this sense it can truly be said that the history of the fair records a significant portion of the state's history.

From the first fair in 1854, the Iowa State Fair has consistently offered a mixture of agricultural exhibits and entertainments, simultaneously promoting economic growth and providing fairgoers an annual carnival. Iowans repeatedly debated the relative place of agriculture and entertainment at the fair. Agricultural exhibits supplied the fair's original purpose, saluting rural life and Iowans' productivity. Entertainments supplied the fair's allure, but they were often eyed warily as promoting the values of urban life, consumption, and leisure. Because Iowa's economy and society were predicated on agriculture, some Iowans fretted that if the fair departed from its agricultural mission their society would lose the very basis of its prosperity. When Iowans debated the place of agriculture, education, and entertainment on the fairgrounds, they were not only urging higher premiums for livestock exhibitors or grousing about a tawdry sideshow but commenting on the past, present, and future of their state.

Iowans regarded the fair as an annual gauge of the state's economic and cultural achievements since the first pioneers crossed into the newly opened territory. The fair was created to promote the state's economic development and annually took the measure of Iowa's progress, tallying the growth of the state's population, economy, and agricultural bounty. The fair told a tale of steady progress, displaying Iowans' accomplishments over the past year and over all the years since white settlement began in 1833.

Alongside its agricultural exhibits and entertainments, the fair also presented its visitors a version of Iowa's history in which sturdy pioneers tamed the frontier and laid the foundation for subsequent agricultural, industrial, and cultural advances. As George C. Dixon noted in his address at the inaugural fair in 1854, the fair itself was an event of historic significance, which marked the beginning of a new era in Iowa's history. The men who founded the Iowa State Agricultural Society and organized the first fair were confident that their society and fair ranked among the most important agents of economic development in the state. They considered the society's annual convention, fair, and yearly volume assessing the state's agriculture important events and documents in the state's heritage, and they meticulously preserved their reports, minutes, and correspondence for posterity.[1]

The fair and its exhibits reaffirmed Iowans' shared heritage by reminding them of their debt to the pioneers who settled the Iowa Territory and broke the prairie, and the fair became a familiar annual ritual that recounted the progress Iowans had made since the pioneer era.

Nearly all of the fair's exhibits were designed to measure and celebrate Iowans' attainments in agriculture, manufacturing, homemaking, crafts, and fine arts: bigger cattle, larger yields, more efficient machinery, nicer homes, finer paintings.

In the late nineteenth century the fair, and Iowans generally, became more retrospective, realizing that they now had accumulated decades of history to recount and preserve. Although virtually every fair offered patrons some lessons in Iowa's history, some fairs were steeped in Iowa's past. The first fair held on the permanent fairgrounds in Des Moines (1886), anniversaries of Iowa's accession as a federal territory (1888 and 1938) and statehood (1896 and 1946), the Closing Century Exposition of 1899, and anniversaries of the first fair (in 1904 and 1929) all emphasized historical themes. These anniversary fairs featured extensive displays of old farm implements, kitchen tools, and whatever other obsolete items exhibitors were willing to haul from their shed or attic to the fair. "Old Settlers" societies held meetings and picnics at the fair, inviting members of the pioneering generation to reminisce about the hardships and satisfactions of life in frontier Iowa. Exhibits of rusted tools, old clothing, and other artifacts of frontier life tangibly reminded visitors how far their state had progressed, eliciting sighs of relief at the relative ease and comfort of contemporary farm life compared to the hardship and privation of the pioneer era. History, in this telling, provided a sure gauge of the advantages of the present and foretold continued advances in the future. As the *Homestead* remarked aptly on the fair's fifty-fifth anniversary celebration in 1909, "Never before were the possibilities of Iowa for the future so blended with the fulfillments of Iowa for the past."[2]

Iowans were not alone in sensing the growing weight of the past. In the late nineteenth century, Americans commonly thought of their nation's history as a tale of progress but began to confront the realization that continued progress was not altogether certain. In 1893 historian Frederick Jackson Turner's influential "frontier thesis" offered the most famous version of this story of progress. According to Turner, the rigors of encountering and subduing the raw wilderness on the frontier had transformed Europeans into Americans and furnished the source of America's individualistic society and democratic political system. Yet, as historians Andrew Cayton and Peter Onuf have perceptively observed, Turner's essay on the advance of civilization belied a nagging fear that the United States was on the brink of a substantial change, or even decline.

The 1890 Census revealed that the frontier era had closed, as the rapid expansion of white settlement had left no uninhabited land for would-be pioneers to settle. Without the safety valve of western land, the nation would invariably become more crowded and more urbanized. If the frontier furnished the source of American individualism and freedom, as Turner asserted, the nation's political system might eventually be endangered. Turner's frontier thesis told a tale of progress, but progress that seemed to be imperiled, or even at an end.[3] Similar worries about the civilization's possible decline existed alongside the paeans to progress on the fairgrounds. The fair's extraordinarily popular grandstand spectacles between 1900 and 1930 relied on well-known historical episodes to tell tales of decline and disaster that brought progress and civilization crashing to a halt in a blaze of pyrotechnic destruction. Instead of unbroken progress, these spectacles offered a cautionary tale in which history was punctuated by catastrophe. Progress, in these spectacles, was not endless, and it sometimes exacted a high price.

In 1929 the Iowa State Fair celebrated its seventy-fifth-anniversary Diamond Jubilee. The Fair Board arranged for extensive exhibits of Iowa's history and erected yet another cardboard volcano to stage a new production of the eruption of Mt. Vesuvius, "Pompeii," billed as the "greatest, most appalling fireworks spectacle of all time."[4] The "Diamond Jubilee" fair capped a decade of remarkable success, setting the fair's all-time attendance record and earning a substantial profit. Afterward, fair secretary Arthur Corey observed that the historical exhibits "attracted the interest of tens of thousands, who would not have cared to see 'just another state fair.'" These visitors did not look at historical exhibits only to shake their heads in wonder at the hardships endured by their pioneer forebears or to marvel at the advent of tractors, telephones, and indoor plumbing. As *Wallaces' Farmer* observed, Iowans were "becoming increasingly conscious and increasingly proud of the history of their state," and so were keenly interested by the fair's extensive historical exhibits.[5]

Only a few weeks after the 1929 fair closed its gates, a disaster all too real befell Americans when the stock market collapsed and the economy plunged into the Great Depression. The state fair was only one of many enterprises ravaged by the Depression, and its attendance and profits plummeted over the next few years. The economic collapse impelled some Iowans to look backward, and to seek reassurance in their history, and inspired the Fair Board to dust off agrarian rhetoric hailing the fair's agricultural mission. In 1932 the Fair Board urged "the rehabilitation of agriculture, and the restoration of the farmer to his rightful place in the economic world," promising that the exhibition would be "a farmer's fair" and "an 'old fashioned' State Fair" that would emphasize agriculture and education, not entertainment. In 1933 the board proclaimed "Back to Fundamentals" as the fair's slogan, insisting that "the time has come for agriculture to return to the proven principles which gave our farm regions their first development and prosperity." Hard times forced midwesterners, like other Americans, to look to their past for the "proven principles" that had led to agricultural progress and economic growth and furnished the fair's original purpose.[6]

By far the fair's most ambitious historical exhibit was mounted during the Great Depression in 1938, to commemorate the centenary of Iowa's accession as a federal territory. Fairgoers passed through a gate that resembled a frontier stockade to discover exhibits of virtually every aspect of Iowa's past, ranging from the development of farm implements and livestock breeding to household decor. The 1938 fair marked the unveiling of regionalist painters Daniel Rhodes's and Howard Johnson's enormous WPA mural of Iowa's history and featured a home-grown historical pageant, "Cavalcade of Iowa." Billed as "the crowning event" of the centennial, "Cavalcade" traced the state's history from the expedition of the French explorers Louis Joliet and Jacques Marquette in the 1670s to the present. Both the mural and "Cavalcade" told a story of economic growth and technological progress, in which civilization advanced steadily through a series of stages from the frontier toward a modern, industrialized society. Fair secretary Arthur Corey declared that "never before has Iowa presented at any one time or place such a comprehensive historical story of her progress and achievements in farming, industry and cultural pursuits."[7]

After a four-year hiatus during World War II, the fair reopened its gates in 1946 to mark the centennial of Iowa's statehood. The searing experience of the war left Iowans only too glad to return to peacetime, and the fair had always been a hopeful emblem of prosperity. The centennial fair chronicled history and heralded progress but seemed for the most part to gaze backward. The fair's main gate, a gleaming Art Deco structure, rose from stockade pickets, symbolizing a century of progress from the slow plodding of horse-drawn wagons to the marvel of commercial airplane travel. Inside the grounds, the fair boasted "100 Years of Iowa History on Parade" and promised to "turn back the calendar to the glory and glamour of frontier Iowa." The fair's historical exhibits included displays of old machinery, modes of travel ranging from oxcarts to airplanes, a full-scale frontier farmstead, pictures of "Livestock Then and Now," and displays of the history of women's fashions and handicrafts. But in the aftermath of world-changing upheaval, some Iowans wondered whether the fair's resumption seemed not so much a return to peace and prosperity but a vestige of a bygone era.[8]

In 1954 the fair marked its centennial with yet another series of historical exhibits and by erecting the Centurlon Spire of Time near the fair's Administration Building. Beneath this spire fair officials buried a time capsule containing artifacts—seeds, letters from prominent Iowans, yearbooks, motion picture footage—to be unearthed at the bicentennial state fair in 2054. The fair's two-hundredth anniversary will undoubtedly inspire a new round of assessments of its history. Iowans in 2054 will doubtless shake their heads in disbelief wonder at the obsolete technology and outmoded styles of the mid-twentieth century as they search for a working movie projector and attempt to thread the celluloid into its sprockets. When the film flickers onto the screen, though, they will also marvel as they recognize how many of the exhibits from the 1954 state fair remain familiar.

Concerns about the fair's relevance and its future viability had persisted throughout the sweeping and accelerating transformations of American life in the late twentieth century. Could the fair keep pace with a society that was being rapidly transformed by interstate highways, television, and air travel? Could the fair remain relevant in a nation remade by postwar affluence and accelerating economic and cultural change? The fair scrambled to keep pace with the cultural upheaval of the 1960s and 1970s, which threatened to make the fair and farm life seem hopelessly dated and square. Teen Town, which featured rock music and vendors of groovy belts, bell bottoms, and "Free Angela Davis" buttons, was constructed to lend the fair a veneer of hipness and attract those on the younger side of the Generation Gap to the fair. But no amount of Day-Glo paint or trippy fonts could lead anyone to mistake the Iowa State Fair for Woodstock. Nagging concerns in the early 1970s that the fair had become out of date and parochial led the Fair Board to imbue the fair with international themes in an effort to attract attendance by offering exhibits never before seen on the fairgrounds: "Discover Mexico," "Discover Hawaii." Displays of Aztec culture and hula dancers, though, seemed oddly out of place on the fairgrounds, and the fair abandoned these exotic themes after only a few years, returning once again to its Iowa roots.

Despite these persistent worries about the fair's relevance, it has proved remarkably adaptable and resilient in the face of sweeping technological, social, and cultural change. Television, like motion pictures before it, posed a challenge to the fair's popularity by competing for Iowans' money and attention. But television has also benefited the fair. Since the early 1970s, Iowa Public Television has broadcast extensive coverage of the fair's events across the state, beaming the fair to those who were unable to attend and enabling those who do attend to see many more exhibits and events than they could possibly take in during a one-day visit. Improved highways and automobiles enabled midwesterners to travel more easily to far-flung destinations during their summer vacation but also made it easier for them to travel to the fair.

Nevertheless, in the late twentieth century, as in the late nineteenth, Iowans continued to worry that the old-time agricultural fair had lost its appeal and its usefulness. Perhaps the state fair had served its purpose and come to the end of its run. The fair's exhibits and entertainments had grown dated, and the fair seemed virtually the same from year to year. Farmers, like city dwellers, now had ample opportunity to obtain information and purchase goods without traveling to Des Moines to visit the fair. Even the iconic fairgrounds, which was added to the National Register of Historic Places in the 1980s, began to seem like a dinosaur and a liability, and the Fair Board estimated that shoring up the fair's aging buildings and infrastructure would cost tens of millions of dollars. Some Iowans advocated pulling up stakes and moving the fair to a new site rather than spending a fortune to preserve the century-old fairgrounds. But others recognized that the fair's long history and its fairgrounds held the key to its enduring appeal. Private donors and corporations contributed money to preserve and improve the fairgrounds, whose layout is as familiar to most Iowans as the streets of their hometown.

Partly as a result of these efforts to save the fairgrounds, the state fair today broadly resembles its predecessors, despite the mind-boggling technological, economic, and social transformations over the past 160 years. A nineteenth-century agriculturist would reel in disbelief at displays of driverless tractors or the sight of farmers checking corn prices on their smartphones, but he would immediately recognize the fairgrounds, livestock judging, and myriad exhibits. He might disapprovingly and avert his gaze as he hurried past the Midway, but he would immediately recognize and feel at home amid the vast crowds, vivid colors, festive sounds, and mouthwatering smells that all remain part of the fair's ineffable ambience.

Acrimonious debates over the place of agriculture and entertainment at the fair have subsided to a faint echo, and both still occupy their rightful place on the fairgrounds. Some fairgoers never set foot on the Midway, while others attend solely for the rides or nighttime concerts. But the sights and sounds of the Midway and the livestock barns are both essential to the fair. The fair is not solely a carnival or a livestock show. The fair unabashedly offers an annual carnival and vacation, yet hearkens to the past and reminds Iowans of agriculture's central place in the state's economy and history. Tellingly, recent fairs have offered both "Non-Stop Fun" and "An Old-Fashioned Good Time."

And something less tangible, yet more essential to the fair, endures as well: the mixture of serious purpose, sociability, and festivity that has lain at the heart of fairs for centuries, and at the heart of the Iowa State Fair since its gates first swung open in 1854. As many fairgoers have observed for over a century and a half, the fair comprises such a bewildering array of exhibits, attractions, activities, sights, and sounds that fairgoers become almost giddy as they enter the grounds—and yet the fair retains a reassuring familiarity. The livestock barns, grandstand, Agriculture Building, and Midway still stand, and fresh-squeezed lemonade and corn dogs invariably prompt Iowans' recollections of fairs past. Iowans go to the fair because it is a cherished annual event, institution, and ritual. The fair has now become a familiar, reassuring annual ritual, a touchstone of stability amid a world of ever-quickening change. Iowans attend the fair for many reasons. Some attend the fair to compete. Some go to buy, and others to sell. Still others go for the country-and-western concerts or the Midway. But most go just to amble about the fairgrounds, to peruse their favorite exhibits, to see what's new, to eat, to people-watch. And they go for another reason as well: because they always have.

Fairs remain popular and enjoyable but have lost some of their former significance. Fairs are still, as Karal Ann Marling writes, "our central cultural institution" in the Midwest, still the metaphorical "heart" of states and counties throughout the region, but few midwesterners today would contend that the annual state or county fair represents the sum total of the region's progress or points the way toward its future development. Instead, the fair has in recent years become an exercise in nostalgia, harkening back to a mythical, bygone era in which a trip to the state fair typified wholesome entertainment and ample evidence of the state's bounty and well-being. Despite its longevity and popularity, the fair's history was seldom free from disagreement over its exhibits and entertainments. Because visitors took seriously the contention that their fair was an exhibition of the state's progress and a harbinger of its prospects, they frequently disagreed over what ought to be permitted on the grounds, what should be kept out, and what it all meant. The annual fair recorded not only the changes that transformed the region and its relationship to the nation but, more important, how midwesterners alternately encouraged, resisted, and understood those changes.[9]

Introduction

Epigraph: Dante M. Pierce, "The State Fair and Its Record of Progress," *Homestead* 68 (1923): 1218.

1.Patricia Schultz, *1,000 Places to See Before You Die: A Traveler's Life List* (New York: Workman Publishing, 2003), 630. For two engaging accounts of the fair's history, see Mary Kay Shanley, *Our State Fair: Iowa's Blue Ribbon Story* (Des Moines: Iowa State Fair Blue Ribbon Foundation, 2000), and Thomas Leslie, *Iowa State Fair: Country Comes to Town* (New York: Princeton Architectural Press, 2007).

2. Karal Ann Marling, *Blue Ribbon: A Social and Pictorial History of the Minnesota State Fair* (St. Paul: Minnesota Historical Society Press, 1990), vii.

3. For thoughtful accounts of the Midwest's place in American history and its neglect by historians, see Andrew R. L Cayton and Susan E. Gray, *The Identity of the American Midwest: Essays on Regional History* (Bloomington: Indiana University Press, 2007); Jon K. Lauck, *The Lost Region: Toward a Revival of Midwestern History* (Iowa City: University of Iowa Press, 2013).

4. Noah Webster, *An American Dictionary of the English Language*; . . . 2 vols. (New York: S. Converse, 1828), "culture."

5. For an introduction to the agrarian myth and its fate in nineteenth-century America, see Leo Marx, *The Machine in the Garden: Technology and the Pastoral Ideal in America* (New York: Oxford University Press, 1964); John F. Kasson, *Civilizing the Machine: Technology and Republican Values in America, 1776–1900* (New York: Grossman, 1976).

6. William Cronon, *Nature's Metropolis: Chicago and the Great West* (New York: W. W. Norton, 1991), 46–54. The classic statement of this conception of frontier development, of course, is Frederick Jackson Turner's 1894 essay, "The Significance of the Frontier in American History," in Clyde Milner, ed. *Major Problems in the History of the American West* (Lexington, MA.: D. C. Heath, 1989).

7. The growth of commercial amusements has been explored by Neil Harris, *Humbug: The Art of P. T. Barnum* (Chicago: University of Chicago Press, 1973); John Kasson, *Amusing the Million: Coney Island at the Turn of the Century* (New

York: Hill and Wang, 1978); Robert Sklar, *Movie-Made America: A Cultural History of American Movies* (New York: Vintage, 1976); Lary May, *Screening Out the Past: The Birth of Mass Culture and the Motion Picture Industry* (New York: Oxford University Press, 1980); Roy Rosenzweig, *Eight Hours for What We Will: Workers and Leisure in an Industrial City, 1870–1920* (New York: Cambridge University Press, 1983); Kathy Peiss, *Cheap Amusements: Working Women and Leisure in Turn-of-the-Century New York* (Philadelphia: Temple University Press, 1986); Lizabeth Cohen, "Encountering Mass Culture at the Grassroots," in her *Making a New Deal: Industrial Workers in Chicago, 1919–1939* (New York: Cambridge University Press, 1990); David Nasaw, *Going Out: The Rise and Fall of Public Amusements* (New York: Basic Books, 1993).

8. On America's consumer economy see Richard Wightman Fox and T. J. Jackson Lears, eds., *The Culture of Consumption: Critical Essays in American History, 1880–1980* (New York: Pantheon, 1983); William Leach, *Land of Desire: Merchants, Money, and the Rise of a New American Culture* (New York: 1993); Jackson Lears, *Fables of Abundance: A Cultural History of American Advertising* (New York: Basic Books, 1994).

9. On efforts to improve life in the countryside, see Michael McGerr, *A Fierce Discontent: The Rise and Fall of the Progressive Movement* (New York: Oxford University Press, 2005), 104–7; for a history of American agriculture in the twentieth century, see R. Douglas Hurt, *Problems of Plenty: The American Farmer in the Twentieth Century* (Chicago: Ivan R. Dee, 2002); Dennis S. Nordin and Roy V. Scott, *From Prairie Farmer to Entrepreneur: The Transformation of Midwestern Agriculture* (Bloomington: Indiana University Press, 2004).

10. James R. Shortridge, *The Middle West: Its Meaning in American Culture* (Lawrence: University Press of Kansas, 1989), 9–10.

Chapter 1

1. Unless otherwise noted, all references to correspondence in this and the following chapters are from the Iowa State Agricultural Society's (ISAS) letter books, held by the State Historical Society of Iowa (SHSI), Des Moines. ISAS, *Report* 1 (1854): 28. Dixon's address runs from pp. 27–48.

2. ISAS, *Report* 1 (1854): passim.

3. Shortridge, *The Middle West*, 3, 5–7, 13–26.

4. On the history of American agricultural fairs, see Wayne Caldwell Neely, *The Agricultural Fair* (New York: Columbia University Press, 1935); on Elkanah Watson, see the same work, esp. 59–71; Donald B. Marti, *To Improve the Soil and the Mind: Agricultural Societies, Journals, and Schools in the Northeastern States, 1791–1865* (Ann Arbor, MI: University Microfilms, 1979), 15–31; Elkanah Watson, *History of the Rise, Progress, and existing Condition of the Western Canals in the State of New-York, . . . together with the Rise, Progress, and existing State of*

Modern Agricultural Societies On the Berkshire System, . . . (Albany, NY: D. Steele, 1820), 159–61.

5. *Statute Laws of the Territory of Iowa,* 1st Sess. (1838–39) (Dubuque: Russell and Reeves, 1839; reprint, Des Moines: Historical Department of Iowa, 1900): 241–43. Myrtle Beinhauer, "The County, District, and State Agricultural Societies of Iowa," *Annals of Iowa,* ser. 3, v. 20, no. 1 (July 1935): 50–69.

6. Earle D. Ross, *Iowa Agriculture* (Iowa City, SHSI, 1951), 24–25; ISAS, *Report* 4 (1857): 410–13.

7. *Fairfield Ledger,* April 11, 1853, October 20, 1853, January 5, 12, 1854; *Iowa Farmer and Horticulturist* 1 (May 22, 1853): 22; Charles J. Fulton, *History of Jefferson County, Iowa* (Chicago: S. J. Clarke, 1914), 259–61.

8. 6th General Assembly (1857), chap. 188.

9. Joshua M. Shaffer (henceforth JMS) to S. L. Smith (Secretary, Cedar County Agricultural Society), November 28, 1868.

10. ISAS, *Report* 3 (1856): 9. Suel Foster, letter in *Iowa Farmer and Agriculturist* 5 (August 1857): 70.

11. JMS et al., Memorial to Iowa Senate and House of Representatives, February 3, 1866; JMS to M. W. Robinson, March 14, 1866.

12. Earle D. Ross, *Iowa Agriculture* (Iowa City, SHSI, 1951), 24–25; ISAS, *Report* 4 (1857): 410–13; John D. Wallace, quoted in ISAS, *Report* 8 (1861): 6.

13. ISAS, *Report* 16 (1869): 29. See also JMS to John McGregor (Jackson County Agricultural Society) December 14, 1865; JMS to J. H. Kelley (Benton County Agricultural Society), December 1, 1866; ISAS, *Report* 24 (1877): 525; John R. Shaffer (henceforth JRS) to A. W. Guernsey (Tama County Agricultural Society), December 30, 1875; ISAS, *Report* 27 (1880): 51. Secretaries of county agricultural societies sometimes found the annual report burdensome. See C. C. Fowler (Des Moines County Agricultural Society) to JRS, October 29, 1883, box C1, f. 1. ISAS, *Report* 5 (1858): 238–39. See also John Scott (Story County Agricultural Society), 12 (1865): 519. On the rivalry between county fairs and the state fair, see J. H. Wallace (Muscatine County Agricultural Society), ISAS, *Report* 4 (1857): 385 (Wallace was secretary of both ISAS and the county society); John W. Irwin, (Mahaska County Agricultural Society), ISAS, *Report* 6 (1859): 325; *History of Linn County* (1878), 402; S. S. Statler (Story County Agricultural Society), ISAS, *Report* 29 (1882): 591.

14. H. S. Hetherington (Northwestern Agricultural Association), ISAS, *Report* 27 (1880): 338. On the advantages of district societies, see Jonas M. Cleland (Woodbury County Agricultural Society), ISAS, *Report* 31 (1884): 688–89. J. Bradley, Cedar Valley District Society, ISAS, *Report* 12 (1865): 383; 20 (1873): 436–37; P. F. Bartle, Central Iowa District Agricultural Society; ISAS, *Report* 11 (1865): 404; 17 (1870): 425; Dixon, *Polk County History Des Moines* (State Register, 1876), 178.

15. On the rivalry between district fairs and the state fair, see the remarks of L. W. Hart (Buchanan County Agricultural Society), ISAS, *Report* 10 (1863): 347; Peter Melendy (Cedar Valley District), 10 (1863): 362, 363–64; J. S. Shepherd (Van Buren County Agricultural Society), 12 (1865): 532; P. F. Bartle, (Central Iowa District Agricultural Society), ISAS, *Report* 13 (1866): 333. ISAS, *Report* 9 (1862): 223; 17 (1870): 30; JMS to John Q. Tufts, 21 February 1871; JMS to John Grinnell, March 24, 1871. H. S. Hetherington (Northwestern Agricultural Association at Dubuque), ISAS, *Report* 27 (1880): 338. On the state society's support for district societies, see JRS to Festus J. Wade, December 8, 1884. See also ISAS, *Report* 25 (1878): 47; 28 (1881): 72; Ross, *Iowa Agriculture*, 86; Beinhauer, "Agricultural Societies," 55.

16. P. F. Bartle, (Central Iowa District Agricultural Society), ISAS, *Report* 11 (1865): 404; Joshua M. Shaffer (Jefferson County Agricultural Society), 9 (1863): 418.

17. On farmers' apathy toward county fairs, see Dr. J. S. Dimmitt (Jones County Agricultural Society), ISAS, *Report* 5 (1858): 282. See also Henry W. Briggs (Davis County Agricultural Society), ISAS, *Report* 4 (1857): 224; C. J. F. Newell (Allamakee County Agricultural Society), 5 (1858): 196; C. Nuckolls (Mills County Agricultural Society), 5 (1858), 324; Robert McKee (Mills County Agricultural Society), 10 (1863): 402; T. T. Pendergraft, (Page County Agricultural Society), 11 (1864): 358; Thomas M. Coleman, (Guthrie County Agricultural Society), 12 (1865): 325; John Scott, (Story County Agricultural Society), 519; E. T. Cole (Davis County Agricultural Society), 13 (1866): 340; report of Cedar County Agricultural Society, 17 (1869): 225; report of Marshal County Agricultural Society, 16 (1869): 274; William C. Chubb (Adams County Agricultural Society), 21 (1874): 319; G. B. Dean (Greene County Agricultural Society), 22 (1875): 372. For complaints about farmers' reluctance to adopt scientific agriculture, see Isaac Kneeland (Lucas County Agricultural Society), ISAS, *Report* 6 (1859): 318; E. T. Cole (Davis County Agricultural Society), 13 (1866): 340, 342–43. See also Z. A. Wellman (Delaware County Agricultural Society), ISAS, *Report* 4 (1857): 237; M. L. Comstock (Des Moines County Agricultural Society), 4 (1857), 243–44; G. F. Kilburn (Adair County Agricultural Society), 11 (1864): 304; T. T. Pendergraft (Page County Agricultural Society), 11 (1864), 358; John Scott, (Story County Agricultural Society), 12 (1865): 519.

18. P. F. Bartle, address to Jasper County Agricultural Society, ISAS, *Report* 12 (1865): 228–40; quotation from 230. See also H. G. Stuart (Lee County Agricultural Society), ISAS, *Report* 4 (1857): 337; John R. Needham (Mahaska County Agricultural Society), 4 (1857): 352; D. W. Kauffman (Van Buren County Agricultural Society), 4 (1857): 420; William Bremner (Marshall County Agricultural Society), ISAS, *Report* 6 (1859): 340; Joshua M. Shaffer (Jefferson County Agricultural Society), ISAS, *Report* 9 (1863): 418–19; B. M. Dewey, address to Chickasaw County Fair, 13 (1866): 314–15; C. W. Magoun, address to Poweshiek County Fair, 18 (1871): esp. 472–73. On the inaugural Montgomery County Fair, see S. S. Merritt, *A History of the County of Montgomery County* (Red Oak: Express Publishing, 1906), 221–22. Farmers' unwillingness to exhibit at their county

fair is discussed by David C. Shaw (Jackson County Agricultural Society), ISAS, *Report* 4 (1857): 283–84; Warren C. Jones (Henry County Agricultural Society, 4 (1857): 249; William H. Kinsman (Pottawattamie County Agricultural Society), 5 (1858): 381; T. G. Gilson (Marion County Agricultural Society), 43 (1896): 414–15; Merritt, *Montgomery County History*, 220.

19. LeGrand Byington (Johnson County Agricultural Society), ISAS, *Report* 4 (1857): 307, 311; report of Des Moines County Agricultural Society, 6 (1859): 230; report of Clayton County Agricultural Society, 16 (1869): 226–27; J. J. Berkey (Fayette County Agricultural Society), 20 (1873): 370; E. A. Walker (Moulton District Agricultural Society), 35 (1888): 483.

20. ISAS, *Report* 23 (1876): 39.

21. *Iowa Farmer and Horticulturist* (1855): 122. JMS to William Stimpson, June 1, 1869.

22. *Daily Gate City* (Keokuk), January 12, 1871.

23. ISAS, *Report* 5 (1858): 21–23; *Weekly Oskaloosa Herald*, January 22, February 26, 1858. JMS to M. W. Robinson, February 19, 1866; JMS to Capt. R. B. Rutledge, March 20, 1866; JMS to James D. Wright, June 27, 1866; JMS to M. W. Robinson, January 3, 1867.

24. *Daily Gate City*, September 15, 1870; *Iowa Homestead and Western Farm Journal* 15 (September 23, 1870): 5. ISAS, *Report* 17 (1870): 39–40. *Cedar Rapids Times*, January 26, 1871.

25. *Cedar Rapids Times*, January 22, 1874.

26. J. R. Shaffer to Oliver Mills, August 26, 1874; *Cedar Rapids Times*, October 7, 1875.

27. On premiums for papers on scientific agriculture, see, for instance, ISAS, *Premium List*, 1857: 11.

28. *Iowa City Republican*, in ISAS, *Report* 7 (1860): 7.

29. Quoted from John Porter's address to the 1881 fair. ISAS, *Report* 28 (1881): 85.

30. ISAS, *Report* (1854): 5.

31. ISAS, *Premium List*, 1857: 8; Wright is quoted in ISAS, *Report* 10 (1864): 62.

32. ISAS, *Report* 3 (1856): 18; ISAS, *Premium List*, 1857: 5. See also *Weekly Oskaloosa Herald*, October 8, 1858; *Burlington Weekly Hawk-Eye*, October 1, 1864.

33. *Iowa Farmer* 1 (1853): 189–91. Secretary John Wallace's words are recounted (and, apparently, paraphrased) in ISAS, *Report* 5 (1858): 28. Clagett is quoted in ISAS, *Report* 1(1854): 23. ISAS, *Report* 10 (1863): 25. The society's annual report often noted approvingly that most fairgoers had submitted cheerfully to the decisions of judges, marshals, and other fair officials. See ISAS, *Report* 4 (1857): 16–17; 11 (1864): 11; 18 (1871): 42.

34. See letters in the State Fair correspondence folder, Dr. Joshua M. Shaffer papers, Manuscripts, State Historical Society of Iowa, Des Moines; ISAS, *Report* 4 (1857): 16; *Dubuque Democratic Herald* September 17, 1863; *Iowa City Republican*, quoted in ISAS, *Report* 7 (1860): 58–59.

35. ISAS, *Report* 15 (1868): 173; ISAS, *Report* 3 (1856): 54; ISAS, *Premium List*, 1857: 9.

36. *Iowa Homestead and Western Farm Journal* 18 (September 19, 1873): 297. See also *Iowa Homestead and Western Farm Journal* 19 (August 28, 1874): 276.

37. *Muscatine Weekly Journal*, October 10, 1857; *Weekly Oskaloosa Herald*, October 8, 1858; ISAS, *Premium List*, 1860: 8–9; 1863: 9; JRS to Oliver Mills, August 26, 1874.

38. ISAS, *Report* 15 (1868): 27.

39. ISAS, *Premium List*, 1857: 9. Welch is quoted in ISAS, *Report* 19 (1872): 184–85.

40. ISAS, *Report* 21 (1874): 92–93.

41. See ISAS, *Report* 27 (1880): passim; 30 (1883): 100, 103.

42. Phil M. Springer to JRS, January 9, 1883, ISAS papers, box Z 1, f. 3. See also JRS to George C. Duffield, December 16, 1882; *Billboard* 8, no. 12 (January 1, 1897): 8.

43. 6th General Assembly (1857): chap. 188, sec. 1; ISAS, *Premium List*, 1860: 6; 1862: 6.

44. Phil M. Springer to JRS, January 9, 1883, ISAS papers, box Z 1, f. 3. See also JRS to George C. Duffield, December 16, 1882; *Billboard* 8, no. 12 (January 1, 1897): 8.

45. ISAS, *Report* 28 (1881): 203–4, 222.

46. JRS to C. W. Norton, November 22, 1884, JRS to N. B. Choate, et al., April 6, 1885; JRS to Choate, May 11, 1885; quotation from JRS to J. D. Brown, April 6, 1885.

47. ISAS, *Report* 31 (1884): 45; JRS to Festus J. Wade (Secretary, Agricultural and Mechanical Association, St. Louis), February 18, 1884; JRS to C. W. Norton, November 22, 1884; JRS to G. H. Grinnell and Son, April 23, 1885; JRS to J. H. Sanders, June 13, 1890.

48. JMS to Joseph Shields, June 26, 1866.

49. ISAS, *Report* 2 (1855): 21; *Iowa Farmer* 4 (September 1856): 120.

50. Oliviez Zunz, *Making America Corporate, 1870–1920* (Chicago: University of Chicago Press, 1999), 149–73.

51. *Iowa Farmer* 5 (November 2, 1857): 152; *Dubuque Democratic Herald*, September 18, 1863.

52. JMS to F. Ellwood Zell and Co., May 29, 1867; JMS to F. J. Upton, September 16, 1867.

53. The illustrations may be found in ISAS, *Report* 17 (1870): 273–312. On the ensuing controversy see ISAS, *Report* 14 (1867): 213–67; JMS to Osborne and Co., July 19, 1871. See also JMS to Aultman, Miller and Co., July 17, 1871.

54. ISAS, *Report* 25 (1878): 10.

55. ISAS, *Report* 23 (1876): 107; 24 (1877): 100; ISAS, *Premium List*, 1883: 46. JRS to N. S. Ketchum, August 27, 1880.

56. JMS to F. Julius LeMoyne, April 2, 1873; JMS to M. W. Robinson, May 1, 1873; JMS to E. B. Shaw, December 4, 1873, JMS to Capt O. R. West, December 23, 1873; JMS to John H. Bacon, February 23, 1874. Mildred Throne, "The Grange in Iowa, 1868–1875," *Iowa Journal of History* 47 (1949): 289–324; D. Sven Nordin, *Rich Harvest: A History of the Grange, 1867–1900* (Jackson: University of Mississippi Press, 1974).

57. *Iowa Homestead* 28 (September 28, 1883): 7; ISAS, *Report* 30 (1883): 271.

58. *Iowa State Register*, April 9, 1885; May 16, May 23, 1885; Polk County Deed Records, Lands, book 142:545, 550; Wesley Redhead and Wife to State of Iowa, ISAS papers, AD VIII, box Z 3 (unclassified); ISAS, *Report* 32 (1885): 61–63; *Iowa State Register*, June 27, 1885; George C. Duffield Diary, June 18, 19, 20, 1885, SHSI, Des Moines; JRS to J. W. Johnson, June 19, 1886.

Chapter 2

1. *Iowa Farmer* 2 (1854): 82, 121.

2. See, for instance, *Iowa Farmer* 2 (1854): 162–65; *Fairfield Ledger*, November 2, 1854; ISAS, *Report* 1 (1854): 19–21.

3. *Iowa Farmer* 2 (1854): 164, 165, 162.

4. *Iowa Farmer* 2 (1854): 165; *Fairfield Ledger*, November 2, 1854.

5. *Iowa Farmer* 2 (1854): 165.

6. *Iowa Farmer* 3 (1855): 47.

7. Mary A. C. Hanford, "A Word on Fairs," *North-Western Farmer* 1 (1856): 271; *The North-West*, in ISAS, *Report* 3 (1856): 33.

8. ISAS, *Report* 3 (1856): 15–17.

9. Businessmen in Dubuque and Burlington sponsored equestrian contests at the fair from 1862 to 1864, and female equestrianism was conducted with the fair's official sanction in 1867 and 1868. *Dubuque Herald*, October 5, 1862; ISAS, *Report* 9 (1862): 143, 204–5; *Dubuque Daily Times*, September 20, 1863; *Burlington Weekly Hawk-Eye*, October 8, 1864; ISAS, *Premium List*, 1867: 20; 1868: 22; ISAS, *Report* 27 (1880): 71, 78.

10. *Iowa State Register*, September 11, 1880.

11. JRS to William Perry, May 5, 1885. JRS to Nellie Curtis, August 13, 1881; JRS

to Nellie Burke, July 2, 1883; NB to JRS, June 29, August 5, 1883, box Sat 1, f. 1.

12. Daniel T. Rodgers, *The Work Ethic in Industrial America, 1850–1920* (Chicago: University of Chicago Press, 1978), 29, 102, 108.

13. ISAS, *Report* 16 (1869): 30–31.

14. *Muscatine Weekly Journal*, October 10, 1857.

15. *Iowa City Republican*, quoted in ISAS, *Report* 7 (1860): 65–66, 74, 65. In 1871 the *Chicago Evening Journal* complained that "six or seven thousand people were allured to the amphitheater all day long, and only a few stragglers were to be seen wandering among the articles on exhibition." *Chicago Evening Journal*, September 16, 1871.

16. ISAS, *Report* 6 (1859): 16.

17. See ISAS, *Premium List*, 1854–1865. In 1865, two trotting matches were allocated premiums of $325 and $200, while thoroughbreds received $188, roadsters $153, and workhorses $182. See also the comments of H. M. Thomson of Scott County at ISAS's convention of 1869. ISAS, *Report* 15 (1868): 177; *Dubuque Democratic Herald*, September 19, 1863; see also *Daily Constitution* (Keokuk), September 19, 1869; *Iowa State Register*, September 4, 1879; 7th General Assembly (1864), chap. 109, sec. 2.

18. ISAS, *Report* 10 (1864): 64.

19. ISAS, *Premium List*, 1874: 19; ISAS, *Report* 19 (1872): 233.

20. JRS to W. Caesar, October 21, 1878. See also JRS to H. G. Cary, December 26, 1878; JRS to A. C. Rogers (Secretary, Clayton County Fair), December 15, 1880.

21. *Proceedings of the Iowa State Agricultural Society in Reference to Securing a Permanent Fair Ground*, [n.d.], pamphlet, SHSI. See also ISAS, *Report* 28 (1881): 229–30.

22. On the society's efforts to classify horses and the growing prominence of trotting and racing, see ISAS, *Premium List*, 1869–1874, esp. 1872: 316; 1874: 122–26; ISAS, *Report* 32 (1885): 589–90; 41 (1894): 181. *Iowa State Register*, September 6, 1881.

23. *Delier v. The Plymouth County Agricultural Society, Iowa Reports* 57 (1881–82): 481–86. See also *Code of Iowa* (1860): secs. 1109–14.

24. *The North-West*, quoted in ISAS, *Report* 3 (1856): 25. See also *Muscatine Weekly Journal*, October 10, 1857. *Iowa City Republican*, quoted in ISAS, *Report* 7 (1860): 59, 67.

25. Recounted in Charles J. Fulton, *History of Jefferson County* (Chicago: S. J. Clarke, 1914), 266.

26. ISAS, *Report* 4 (1857): 17–18.

27. JMS diary, September 8, 1871, Manuscripts, SHSI, Des Moines; *Cedar Rapids Times*, September 12, 1872.

28. *Daily Constitution* (Keokuk), September 18, 1869; George A. Schafer (of G. A. S. and Co., Manufacturers of True Pharmaceuticals) to JRS, September 11, 1883, ISAS papers, box Sat 1, f. 1.

29. JMS to John Scott, August 14, 1873. See also JMS to Edwin Smith, August 15, 1873; JMS to William B. Leach, August 15, 1873; JMS to William Paist, August 16, 1873; ISAS, *Report* 19 (1872): 233.

30. *Dubuque Democratic Herald*, September 17, 1863; *Daily Constitution* (Keokuk), September 17, 18, 1869; *Cedar Rapids Times*, September 18, 1873.

31. *Daily Gate City*, September 26, 28, 29, 1874, September 24, 26, 28, 1875; "Fair Week '84," *Iowa State Register*, September 7, 1884. See also "Crooks Coming to Town for State Fair Week," *Des Moines Leader*, September 6, 1896.

32. The first clear-cut receipt from sideshows, for $474, appears in the society's financial records of 1865. ISAS, *Report* 11 (1864): 63.

33. ISAS, *Report* 11 (1864): 64–65.

34. ISAS, *Report* 16 (1869): 71; *Report* 17 (1870): 81–82; *Daily Gate City*, September 11, 1870; JMS to Hon. C. F. Davis, August 27, 1870.

35. JMS to Col. Milo Smith, August 25, 1871; JMS to Col. S. F. Spofford, August 26, 1871; *Iowa Homestead and Western Farm Journal* 18 (September 19, 1873): 297.

36. ISAS, which had enjoyed annual surpluses of roughly $7,000 in recent years, was unable to pay its premiums from 1876 through 1879 and remained afloat only by borrowing more than $15,000. See ISAS, *Report* 22 (1875): 30–32½, 101–4; 24 (1877): 531–33; 25 (1878): 39–40; 26 (1879): 6–7, 106.

37. In 1874 the society's appropriation was cut from $2,000 to $1,000; in 1875, and from 1878 through 1887, it received no state money whatsoever. The fair's receipts also dropped sharply between 1874 and 1878. ISAS, *Report* 19 (1872): 106. JRS to J. J. Snouffer, July 12, 1875. See also JRS to Hon. C. F. Davis, August 14, 1875.

38. *Daily Gate City*, October 1, 1875, Fair Supplement, [1]. *Western Farm Journal* 20 (October 8, 1875): 313. See also JRS to S. E. Laird, August 17, 1875; JRS to T. Beldin, August 24, 1875; JRS to T. H. Clutter, August 24, 1875.

39. ISAS, *Report* 22 (1875): 202–3; 23 (1876): 98; JRS to E. S. Fonda, January 18, 1876; *Cedar Rapids Times*, September 27, 1877; JRS to Oliver Mills, June 3, 1878; JRS to ISAS Board, June 3, 1878; JRS to S. F. Spofford, June 21, 1878; ISAS, *Report* 25 (1878): 164.

40. On gambling, see Karen Halttunen, *Confidence Men and Painted Women: A Study of Middle-Class Culture in America, 1830–1870* (New Haven, CT: Yale University Press, 1982), 16–20; Ann Vincent Fabian, *Card Sharps, Dream Books and Bucket Shops* (Ithaca, NY: Cornell University Press, 1990); Jackson Lears, *Something for Nothing: Luck in America* (NY: Viking, 2003).

41. ISAS, *Report* 24 (1877): 530.

42. JRS to George E. Bryant (Secretary, Wisconsin Board of Agriculture), July 17, 1879.

43. ISAS, *Report* 27 (1880): 203–4; 28 (1881): 75; JRS to T. B. Hotchkiss, July 12, 1881. On sideshows at the 1881 fair, see *Iowa State Register*, September 6, 1881.

44. JRS to J. F. Merry, July 4, 1881.

45. JRS to William T. Smith, April 14, 1885.

46. ISAS, *Report* 28 (1881): 86. Porter's worries may well have been prompted by George Beard's work on neurasthenia, or "American nervousness." George Beard, *A Practical Treatise on Nervous Exhaustion (Neurasthenia)* (New York: William Wood, 1880).

47. ISAS, *Report* 28 (1881): 86–87.

48. *Iowa State Register*, September 8, 1882.

49. ISAS, *Report* 29 (1882): 339; *Iowa State Register*, September 4, 1883.

50. Nellie Burke to JRS, January 1, 1883, ISAS papers, box Sat 1, f. 1; Harry Stadden to JRS, July 14, 1883, box Sat 1, f. 2; George L. Davenport (of the Sac and Fox Agency, U. S. Indian Service) to E. F. Brockway, July 21, 1883, box Sat 1, f. 1; JRS to William T. Smith, May 20, 1884; JRS to J. J. Snouffer, May 20, 1884; Frank E. Yates to JRS, May 31, 1884, AD I, II; Gomes and Jewell, Practical Aeronautical Engineers, to JRS, June 4, June 13, June 17, 1884, box Sat 1, f. 3; JRS to L. C. Baldwin, June 18, 1884; ISAS, *Report* 31 (1884): 53–56. J. J. Snouffer to JRS, August 18, 1883, ISAS papers, box Sz 1, f. 6; William T. Smith to Sitting Bull, Gall, Crow King, and Rain-in-the-Face, August 24, 1883. The society paid the Sioux chiefs $800 for appearing at the fair. ISAS, *Report* 30 (1883): 55, 58, 307. Smith's remarks may be found on p. 71.

51. "State Fair Scenes," *Persinger Times* (Des Moines), September 11, 1886.

52. These speeches are reprinted in ISAS, *Report* 33 (1886): 89–101, and in the *Iowa State Register*, September 8, 1886.

53. ISAS, *Report* 33 (1886): 96–97.

54. *Persinger Times* (Des Moines), September 11, 1886.

55. *Iowa State Register*, August 29, September 1, 1888; ISAS, *Report* 35 (1888): 604. The society's most outspoken foe of amusements, George Wright, objected that President H. C. Wheeler's remarks "did not go far enough," and called for the total abolition of sideshows. ISAS, *Report* 35 (1888): 618.

56. ISAS, *Report* 36 (1889): 613.

57. ISAS, *Report* 36 (1889): 613. At the 1890 fair, no gambling or "indecent institutions" were allowed. See *Iowa State Register*, August 30, 1890.

58. *Iowa State Register*, September 5, 1889; ISAS, *Report* 36 (1889): 103.

59. JRS to T. L. Newton (Secretary, Wisconsin State Fair), March 19, 1888; JRS to John A. Evans, July 12, 1889. See also John Hayes to JRS, ISAS papers, February 1, 1889, box Sz 2, f. 13, ; JRS to Hayes, March 7, 1889. JRS to Edson Gaylord, June 5, 1890; JRS to Jonathan Periam, June 14, 1888; JRS to Messrs. Hammack and Allen, June 12, 1889; ISAS, *Report* 39 (1892): 580–81. See also *Homestead* 37 (1892): 876.

60. *Homestead* 48 (1903): 29.

61. *Iowa State Register*, September 3, 1892; ISAS, *Report* 39 (1892): 110, 531–32; JRS to N. S. Ketchum [March 1, 1893]. On the society's decision to host a fair in 1893, see *Iowa State Register*, September 2, 1893; see also *Iowa State Register*, August 30, 1893, September 6, 1893; ISAS, *Report* 40 (1893): 126–27.

62. See the *Farmer's Tribune*, August 30, 1893; *Iowa State Register*, September 5, 1894. As for Shaffer's dim view of the Farmer's Alliance, he predicted in 1887 that "if that organization goes into politics there will be ruptures, you mark my word for it." JRS to James Smith, April 8, 1887. On the history and comparative weakness of Populism in Iowa, see Jeffrey Ostler, *Prairie Populism: The Fate of Agrarian Radicalism in Kansas, Nebraska, and Iowa, 1880–1892* (Lawrence: University Press of Kansas, 1993).

63. ISAS, *Report* 40 (1893): 114; JRS to John A. Evans, July 19, 1893; JRS to JMS, July 25, 1893; JRS to O. E. Skiff, July 27, 1893; JRS to Evans, August 5, 1893, July 7, 1893.

64. ISAS, *Report* 40 (1893): 120–22; 39 (1892): 519; *Iowa State Register*, September 2, 1893; JRS to J. W. McMullin, March 29, 1893; JRS to John A. Evans, April 5, 1893.

65. ISAS, *Report* 40 (1893): 114; JRS to John A. Evans, 19 July 1893; JRS to JMS, July 25, 1893; JRS to O. E. Skiff, July 27, 1893; JRS to Evans, August 5, 1893; JRS to Evans, August 7, 1893. ISAS, *Report* 40 (1893): 124, 133–34; *Iowa State Register*, August 31, 1893; JRS to J. W. McMullin, July 15, 1893; JRS to McMullin, July 19, 1893.

66. *Iowa State Register*, August 30, 1893. In 1892 the society equipped the fairgrounds with electric lighting in order to host nighttime entertainments and stem the exodus of fairgoers into the city each evening. ISAS, *Report* 36 (1889): 635; 37 (1890): 653; 38 (1891): 132. On the 1893 fair's disappointing receipts, see *Iowa State Register*, September 8 and 9, 1893. The fair's receipts for 1893 totaled $25,435, less than half the amount of each of the previous four years. As a result, the society, which had already borrowed $8,000 in 1892, was forced to borrow an additional $22,000 in 1893, and it closed the year with debts totaling $24,987. Shaffer's apologetic letters to exhibitors may be found in Letters Books 46–47. ISAS, *Report* 40 (1893): 132, 472. For Shaffer's explanation of the fair's failure, see ISAS, *Report* 40 (1893): 11–13. On his attempt to shore up his support and retain his job as secretary, see JRS to Robert W. Furnas (Secretary, Nebraska State Fair), December 4, 1893; JRS to I. J. Swain, December 8, 1893; JRS to A. B. Hosbrook, December 9, 1893. In March the legislature stated that "it would be a

public calamity" if the fair were discontinued, and appropriated $10,000 for the society in both 1894 and 1895. Still in the red, ISAS borrowed another $15,700 to pay its debts. *Homestead* 39 (1894): 200; 25th General Assembly (1894), chap. 137.

67. ISAS, *Report* 40 (1893): 459; P. L. Fowler to Charles F. Kennedy (Secretary, Indiana State Fair), May 15, 1894; Fowler to Daniel Sheehan, April 18, 1894.

68. *Iowa State Register*, September 8, 1894.

69. *Iowa State Register*, September 8, 1894: 4. See also *Homestead* 42 (1897): 797.

70. James R. Shortridge states that the demise of Middle Western self-confidence "can be dated rather precisely at 1920." Shortridge, *The Middle West*, 39. See also 9, 37–38, 39–57.

Chapter 3

1. *Des Moines Register and Leader*, August 27, 1906.

2. Herbert Quick, "The Women on the Farms," *Good Housekeeping* 57 (1913): 427.

3. On the lives of farm youth, see Pamela Riney-Kehrberg, *Childhood on the Farm: Work, Play, and Coming of Age in the Midwest* (Lawrence: University Press of Kansas, 2005). On farm women, see Katherine Jellison, *Entitled to Power: Farm Women and Technology, 1913–1963* (Chapel Hill: University of North Carolina Press, 1993); Mary Neth, *Preserving the Family Farm: Women, Community, and the Foundations of Agribusiness in the Midwest, 1900–1940* (Baltimore: Johns Hopkins University Press, 1995).

4. Jackson Lears, *Rebirth of a Nation* (New York: Harper, 2009), 135–36.

5. "Better County Fairs For Iowa," *Homestead* 66 (1921): 2161; "The Spirit of the Fair," *Scribner's* 56 (October 1914): 552–53.

6. Quick, "Women on the Farms," 427. *Iowa Yearbook of Agriculture* 14 (1913): 220; 18 (1917): 126–27; "Better County Fairs for Iowa," *Homestead* 66 (1921): 2161.

7. Theodore Roosevelt to Liberty Hyde Bailey, 10 August 1908, in *Report of the Country Life Commission* (New York: 1911; reprint, New York: Sturgis and Walton, 1917), 44. On the Country Life movement, see Michael McGerr, *A Fierce Discontent: The Rise and Fall of the Progressive Movement in America, 1870–1920* (New York: Free Press/Simon and Schuster, 2003), 104–107; David Danbom, *The Resisted Revolution: Urban America and the Industrialization of Agriculture, 1900–1930* (Ames: Iowa State University Press, 1979), 43–44.

8. *Report of the Country Life Commission*, 103–6.

9. John Hamilton, *Agricultural Fair Associations and Their Utilization in Agricultural Education and Improvement*, U. S. Department of Agriculture, Office of Experiment Stations, Circular 109 (Washington, DC: Government Printing Office, 1911).

10. *Homestead* 52 (1907): 964; 61 (1916): 1529; 58 (1913): 1260. On the attitudes of fair men, see C. E. Cameron, "The Educational Value of the State Fair," *Billboard* 24, no. 50 (December 14, 1912): 38; "The Modern Fair," *Greater Iowa*, May 15, 1909: 3; "Educational Value of Fairs," *Greater Iowa*, June 15, 1910: 3; C. E. Cameron, "Substantial Benefits Drawn from the State Fair," *Greater Iowa*, August 17, 1912: 1; "Iowa State Fair Greater Teacher Than University," *Greater Iowa*, July 1, 1913: 1. *Greater Iowa*, first published in 1913, was the fair's newspaper and press release and was distributed to journalists, stock breeders, and farmers to publicize the fair.

11. *Northwest Farmer*, quoted in ISAS, *Report* 6 (1859): 70.

12. ISAS, *Report* 24 (1877): 93–94.

13. JRS to Mary J. Aldrich, October 13, 1885; *Woman's Standard*, September 1886: October 1886: 5; *Iowa State Register*, September 4, 1886. On Woman's Day at the fair, see *Woman's Standard*, September 1891 and October 1891; *Iowa State Register*, September 3, 1891; ISAS, *Report* 38 (1891): 139; *Iowa State Register*, September 4, 1892; P. L. Fowler to Mrs. C. V. Weaver, March 14, 1895; Fowler to John A. Evans, March 25, 1895; Fowler to Mrs. M. E. McGonigal, May 6, 1895. Participants in Woman's Day activities at the 1895 fair included the P.E.O., WCTU, YWCA, nonpartisan WCTU, Iowa Federation of Women's Clubs, Daughters of Rebekah, I.O.O.F., IWSA, King's Daughters' Union, and D.A.R. ISAS, *Report* 42 (1895): 287; *Iowa State Register*, September 11, 1895.

14. *Iowa Yearbook of Agriculture* 4 (1903): 78, 95; 5 (1904): 97, 101–2, 104; 7 (1906): 468.

15. *Iowa Yearbook of Agriculture* 10 (1909): 269; 11 (1910): 298. O. C. Simonds, *Landscape-Gardening* (New York: Macmillan, 1920). Simonds's ambitious plan for the fairgrounds is held by the Iowa State Fair Board, Iowa State Fairgrounds, Des Moines.

16. *Homestead* 57 (1912): 1587; *Greater Iowa*, May 1913: 8; August 1914: 1, 3.

17. Clarke's words are paraphrased in *The Des Moines Register and Leader*, August 29, 1914; "Farm Women at the State Fair," *Homestead* 64 (1919): 2153.

18. *Iowa Yearbook of Agriculture* 15 (1914): 58. *The Des Moines Register and Leader*, August 29, 1914.

19. *Iowa Homestead and Western Farm Journal* 17 (1872): 308–9; *Cedar Rapids Republican*, August 8, 1876, reprinted in *Iowa State Register*, August 11, 1876; *Cedar Rapids Times*, September 21, 1876, September 20, 1877.

20. ISAS, *Report* 30 (1883): 71; 31 (1884): 65.

21. On the disturbing history of eugenics in America, see Edwin Black, *War Against the Weak: Eugenics and America's Campaign to Create a Master Race* (New York: Dialog Press, 2012); Paul A. Lombardo (ed.), *A Century of Eugenics in America: From the Indiana Experiment to the Human Genome Era* (Bloomington:

Indiana University Press, 2011); Robert W. Rydell, "'Fitter Families for Future Firesides:' Eugenics Exhibits at American Fairs and Museums Between the World Wars," in Robert W. Rydell, *All the World's a Fair* (Chicago: University of Chicago Press, 1993), 38–59.

22. *The Des Moines Register and Leader*, August 22, 1912, Fair magazine section, 2; *Iowa Yearbook of Agriculture* 13 (1912): 449; *Homestead* 59 (1914): 1574.

23. On the origins of the baby judging, see *Iowa Yearbook of Agriculture* 13 (1912): 450. In 1911 the baby beauty contest attracted 388 entries. *The Des Moines Register and Leader*, August 30, 1911; Iowa Congress of Parents and Teachers, *The First Fifty Years* (Davenport: Iowa Congress of Parents and Teachers, 1950), 61–62.

24. *Iowa Yearbook of Agriculture* 13 (1912): 248, 285; *Homestead* 57 (1912): 1586.

25. *Wallaces' Farmer* 37 (1912): 1257; *Homestead* 57 (1912): 1587; *The Des Moines Register*, August 2, 1922, Fair magazine section, 2. On the growing popularity of the contest, see Mary T. Watts to *Homestead* 59 (1914): 1488; *The Des Moines Register*, August 29, 1920, August 23, 1923; Iowa State Fair Board, *Official Catalog* (1912): 65; (1913): 78.

26. *Greater Iowa*, July 1912: 3; July 1914: 1, 5; Pownall, "Fitter Families," 431. In 1924 Watts addressed the convention of the International Association of Fairs and Expositions on "fitter families." *Billboard* 36, no. 50 (December 13, 1924): 203. *The Des Moines Register and Leader*, August 23, 1913. See also *Greater Iowa*, October 1913: 8.

27. *Des Moines Register*, August 25, 1929; *The Des Moines Register and Leader*, August 29, 1911. On the evaluation of the babies' parents, see *The Des Moines Register and Leader*, August 24, 1913.

28. *The Des Moines Register*, August 28, 1927, August 30, 1931.

29. *The Des Moines Register*, August 24, 1924, August 8, 1925, August 26, 1926, August 30, 1931.

30. Mary T. Watts, "Better Babies on Our Farms," *Homestead* 59 (1914): 1223. See also *Wallaces' Farmer* 39 (1914): 1234; *The Des Moines Register and Leader*, September 4, 1915.

31. *The Des Moines Register and Leader*, August 25, August 27, 1912; "Give the Farm Babies a Chance," *Homestead* 60 (1915): 1824; see also *Homestead* 59 (1914): 1193. Rural babies were a minority of contest entrants until 1938. *The Des Moines Register*, August 26, 1938.

32. *Wallaces' Farmer* 48 (1923): 1188, quoted in *Proceedings*, State Agricultural Convention 1 (1923): 35.

33. *The Des Moines Register*, August 26, 1924; *Homestead* 69 (1924): 1349.

34. *The Des Moines Register*, September 3, 1926.

35. *The Des Moines Register*, August 26, 29, 31, 1926.

36. On the judges' admiration for plump farm maids, see *The Des Moines Register*, August 31, 1926. The *Register*'s remarks on Alberta Hoppe are quoted in *Greater Iowa*, October 1935: 8; Iowa State Fair Board, *Annual Report*, 1935: 19. *The Des Moines Register*, September 3, 1926. *The Des Moines Register*, September 1, 3, 1937.

37. *The Des Moines Register*, August 28, 1929, August 30, 1933, September 2, 1936; Iowa State Fair Board, *Annual Report*, 1936: 29.

38. On the origins of the 4-H, see Neely, *The Agricultural Fair*, 135–45.

39. *Iowa Farmer and Horticulturist* 5 (1857): 151; on the children's department at the fair, see ISAS, *Report* 22 (1875), 23 (1876), 24 (1877); *Bushnell's Des Moines City Directory* 7 (1879): 51; *Iowa State Register*, September 2,1879; ISAS, *Report* 29 (1882): 336; O. H. Benson to J. C. Simpson, September 6, 1909, ISAS papers, box Se 15, f. 141; *Iowa Yearbook of Agriculture* 10 (1909): 262–65. On efforts to persuade farmers to bring their wives and children to the fair, see "Come to the Fair," *Iowa State Register*, September 6, 1887; *Homestead* 36 (1891): 796; "Give the Boys and Girls a Chance to Attend the Fair," *Wallaces' Farmer*, quoted in *Iowa Yearbook of Agriculture* 5 (1904): 562–63; "The Farm Boy and the State Fairs," *Wallaces' Farmer* 30 (September 8, 1905): 1039; "Attending the State Fair," *Wallaces' Farmer* 36 (August 18, 1911): 1137.

40. *Homestead* 69 (1924): 1299.

41. See, for example, the State Board of Agriculture, *Official Catalog*, 1926: 43–47; Josephine Bakke's address to the Iowa Fair Managers Convention (1925): 373–74.

42. Score sheets may be found in the Iowa 4-H Girls' Club Historian's books, Department of Special Collections, Iowa State University Library, series 16/3/4, boxes 41–48.

43. On the growth of the 4-H, see *Homestead* 65 (1920): 2521–22; Josephine Arnquist to county 4-H agents, July 22, 1922, in Iowa 4-H Girls' Club Historian's Book, Department of Special Collections, Iowa State University Library, series 16/3/4, box 41; *Greater Iowa*, December 1922: 3; *The Des Moines Register*, August 26, 1926, August 27, 1927, August 28, 1931. On the regimentation of 4-H activities at the fair, see *The Des Moines Register*, August 28, 1928; Iowa 4-H Girls' Club Historian's Book, 1934, chap. 10, Department of Special Collections, Iowa State University Library, series 16/3/4, box 44.

44. *Homestead*, September 6, 1928, quoted in *Greater Iowa*, October 1928: 2. In 1934 the Dean of Agriculture at Iowa State College, H. H. Kildee, declared that promoting 4-H club work was an "opportunity of greatest importance to our fairs at this time." *Greater Iowa*, May 1934: 8; *The Des Moines Register*, August 23, 1936. *Wallaces' Farmer*, quoted in Iowa State Fair Board, *Annual Report*, 1924: 36; D. F. Malin, "Iowa's Future Farmers at the Fair," *Wallaces' Farmer* 50 (1925): 1126.

45. *Greater Iowa*, May 1934: 2. See also *Greater Iowa*, October 1934: 4.

46. *The Des Moines Register*, August 19, 25, 1928, August 27, 1932, August 31, 1933, August 19, 1934 (fair advertising section, p. 25), August 31, 1939; *Homestead* 74 (1929): 1282. *The Des Moines Register*, September 1, 1932.

47. *The Des Moines Register*, August 16, 1934.

48. *The Des Moines Register*, August 16, 1934.

49. "The Fair from a Woman's Viewpoint," *Wallaces' Farmer*, September 9, 1927, in Iowa State Fair Board, *Annual Report*, 1927: 25.

50. *Greater Iowa*, August 1914: 5.

51. *Greater Iowa*, July-August 1922: 5; *The Des Moines Register*, August, 24, 27, 1922 (state fair section, p. 11). See also Josephine Wylie, "What Interests Women at the Fair?" and "Home Makers at the Fair," *Wallaces' Farmer*, September 7, 1928, reprinted in Iowa State Fair Board, *Annual Report*, 1928: 47–53. On the history of home economics and its connection to consumer culture, see Carolyn M. Goldstein, *Creating Consumers: Home Economists in Twentieth-Century America* (Chapel Hill: University of North Carolina Press, 2012); Megan Elias, *Stir It Up: Home Economics in American Culture* (Philadelphia: University of Pennsylvania Press, 2010); Sarah Stage and Virginia B. Vincenti, eds. *Rethinking Home Economics: Women and the History of a Profession* (Ithaca, NY: Cornell University Press, 1997).

52. *Wallaces' Farmer* 56 (1931): 1020; "Women's Building Housed New and Unusual Events," *Wallaces' Farmer*, reprinted in *Greater Iowa*, October 1931: 6.

53. *Homestead* 68 (1923): 1337–38.

54. *Homestead* 73 (1928): 1357–59.

55. *Wallaces' Farmer* 56 (1931): 1020.

56. *Wallaces' Farmer* 56 (1931): 1020.

57. *Homestead* 67 (1922): 1389; *The Des Moines Register*, September 2, 1927. On household technology, see Ruth Schwarz Cowan, *More Work for Mother: The Ironies of Household Technology from the Open Hearth to the Microwave* (New York: Basic Books, 1983), 187–88; Susan Strasser, *Never Done: A History of Housework* (New York: Henry Holt, 2000); Janice Rutherford, *Selling Mrs. Consumer: Christine Frederick and the Rise of Household Efficiency* (Athens: University of Georgia Press, 2003).

58. *Wallaces' Farmer*, quoted in Iowa State Fair Board, *Annual Report*, 1926: 47.

59. Iowa State Fair Board, *Annual Report*, 1926: 110. On women as consumers, see Regina Lee Blaszczyk, *American Consumer Society, 1865–2005: From Hearth to HDTV* (Wheeling, IL.: Harlan Davidson, 2009).

60. See, for example, *Des Moines Leader*, August 26, 1900, in which the Younker Brothers department store billed itself as "Headquarters for State Fair Visitors."

61. Recognizing that some rural Iowans might feel bewildered in Des Moines's stores, *Wallaces' Farmer* in 1913 published shopping tips for fairgoers. *Wallaces' Farmer* 38 (1913): 1102. *Greater Iowa*, August 1936: 7; *The Des Moines Register*, August 23, 1936; *Wallaces' Farmer*, quoted in *Greater Iowa*, October 1937: 4.

62. State Agricultural Convention, *Proceedings*, 1926: 108; "A Fair for All," State Agricultural Convention, *Proceedings*, 1938: 178.

63. *Homestead* 66 (1921): 1378, 1341. *The Des Moines Register*, August 31, 1921. The Country Theater proved popular, and the State Board of Agriculture agreed to contribute to its support in 1922. *Iowa Yearbook of Agriculture* 23 (1922): 8. The Little Country Theater movement began with the first "Little Theater," begun in 1900 by the Hull House Players in Chicago. See Marjorie Patten, *The Arts Workshop of Rural America* (New York: 1937), chap. 5; Clarence A. Perry, *The Work of the Little Theaters*, (New York: Russell Sage Foundation, 1933).

64. "Selected Plays for Amateur Production in Township and Community Meetings," Shattuck papers, f. 1/5.

65. *The Des Moines Register*, September 2, 1925, August 26, 1926.

66. These plays were among those presented at the 1922 and 1923 fairs. *Greater Iowa*, July 1922: 5; August 1923: 12; *The Des Moines Register*, August 29, 1922. Handbills and scripts may be found in the Shattuck papers and 4-H Historian's Books, Department of Special Collections, Iowa State University. For short descriptions of plays, see "Selected Plays," Shattuck papers, f. 1/5.

67. Mignon Quaw, *What Every Woman Knows*, 1–2. Quaw, a prolific Extension Service playwright, lectured on the beneficial effects of drama at the 1929 fair. *The Des Moines Register*, August 25, 1929. On some rural families' misgivings about the value of home economics extension, see Danbom, *Resisted Revolution*, 86–94.

68. Quaw, *What Every Woman Knows*, 8.

69. These "rural" plays were all Extension Service creations, which extolled eating a balanced diet, maintaining household account books, and holding occasional family sing-alongs. "Plays at the State Fair," *Wallaces' Farmer* 48 (September 7, 1923): 1188.

70. "Report of the [Iowa State College] State Fair Committee," 1923. Arthur L. Bakke papers. Department of Special Collections, Iowa State University, series 13/5, f. 2/8.

71. Shattuck to Mrs. H. E. Scott, November 14, 1923. Shattuck papers, f. 1/3; *Greater Iowa*, August 1936: 5.

72. *Greater Iowa*, August 1936: 5. See also "Farmers Go to Bed with the Chickens? Oh No, Not in Iowa!" *The Des Moines Register*, August 27, 1933; *Greater Iowa*, July 1937: 2; August 1939: 3; October 1939: 5; August 1941: 5. "The Upward Trend," *Billboard* 40, no. 12 (March 24, 1928): 52; "Drama at the Fairs," *Billboard* 39, no. 30 (July 23, 1927): 40.

73. *Variety* 119, no. 5 (July 17, 1935): 1; Mrs. Eugene Cutler, "A Fair for All," State Agricultural Convention (1938): 178–79.

Chapter 4

The title of this chapter is taken from a 1928 advertisement for the fair published in *The Des Moines Register*.

1. "Evolution, Not Revolution in Fairs and Fair Management Is the Thing To Be Desired," *Billboard* 27, no. 13 (March 27, 1915): 22.

2. On the carnivalesque, see Lears, *Fables of Abundance* and *Rebirth of a Nation*.

3. On mass entertainment in the late nineteenth and early twentieth centuries, see Harris, *Humbug*; Roy Rosenzweig, *Eight Hours for What We Will: Workers and Leisure in an Industrial City, 1870–1920* (New York: Cambridge University Press, 1983); John Kasson, *Amusing the Million: Coney Island at the Turn of the Century* (New York: Hill and Wang, 1978); Kathy Peiss, *Cheap Amusements: Working Women and Leisure in Turn-of-the-Century New York* (Philadelphia: Temple University Press, 1986); Lary May, *Screening Out the Past: The Birth of Mass Culture and the Motion Picture Industry* (New York: Oxford University Press, 1980); David Nasaw, *Going Out: The Rise and Fall of Public Amusements* (New York: Basic Books, 1993).

4. The Iowa State Fair participated in many different circuits with various state and county fairs over the years. G. W. Harvey, "The Modern Agricultural Fair," *Iowa Yearbook of Agriculture* 6 (1905): 999; C. N. Cosgrove (Secretary, Minnesota State Fair) to John C. Simpson, January 5, 1909; JCS to Cosgrove, January 12, 1909, State Board of Agriculture papers (henceforth SBA papers), box Sat 6, f. 58; Jouett Shouse (Secretary, Blue Grass Fair Association) to JCS, February 11, 1909, SBA papers, box Sz 14, f. 127; letters between B. E. Gregory (Superintendent of Publicity and Amusements, Minnesota State Fair) and JCS, 1909, SBA papers, box Sat 6, f. 59. S. M. Yeaman to JRS, March 1, 1884, ISAS papers, box Z 1, f. 4; Festus J. Wade (Secretary, International Association of Fairs and Expositions) to JRS, December 10, 1884, ISAS papers, Box Z 1, f. 4; JRS to Wade, 29 January 1885; ISAS, *Report* 32 (1885): 112.

5. P. L. Fowler to E. W. Randall (Secretary, Minnesota State Fair), April 9, 1897. See JRS to J. B. Peck, July 10, 1888; JRS to W. G. Garrard (Secretary, Illinois State Fair), May 11, 1891.

6. *Billboard* 10, no. 12 (December 1, 1898): 26.

7. *Billboard* 12, no. 2 (December 1, 1899): 16.

8. On the growth of midways, see Neely, *The Agricultural Fair*, 201–12; on the Midway at the Minnesota State Fair, see Marling, *Blue Ribbon*, 185–94.

9. PLF to F. N. Chase, July 19, 1894; PLF to John A. Evans, July 21, 1894; PLF to F. N. Chase, July 21, 1894; PLF to William F. Cross (Secretary, Minnesota State

Fair), [July 9 or 10, 1894]; *Iowa State Register,* September 1, 1894; ISAS, *Report* 41 (1894): 173–74. When the 1894 fair ended, the society was still $16,000 in debt. ISAS, *Report* 41 (1894): 14.

10. PLF to George W. Franklin, December 18, 1894.

11. PLF to John A. Evans, August 3, 1895; *Iowa State Register,* August 28, 1895.

12. *Iowa State Register,* August 28, 1895.

13. ISAS, *Report* 43 (1896): 8–92.

14. ISAS, *Report* 44 (1897): 117.

15. ISAS, *Report* 45 (1898): 121. See also ISAS, *Report* 44 (1897): 114, 117–18; 45 (1898): 18; 46 (1899): 112–13; *Homestead* 44 (1899): 989–93.

16. G. H. Van Houten to W. F. Harriman, March 2, 1899; GHV to ISAS members, March 7, 1899; GHV to L. H. Pickard, March 7, 1899; *Des Moines Leader,* August 28, 1899. The fair was a financial success, and the society closed the year with more than $16,000 in its treasury, its first significant surplus since 1891.

17. Fowler's remarks on the fair's diminished importance may be found in ISAS, *Report* 44 (1897): 117. 28th General Assembly (1900), chap. 58, secs. 1–8. This bill was approved March 21, 1900. The society's final report and minutes are included in the *Iowa Yearbook of Agriculture* 1 (1900): 174–90.

18. *Iowa Yearbook of Agriculture* 3 (1902): 96, 112; 4 (1903): 75. See also "A Look Back at the Fairs," *Homestead* 47 (1902): 327; 48 (1903): 511; "State Fair Fakers Must Go," *Homestead* 48 (1903): 591.

19. *The Des Moines Register and Leader,* August 23, 1903; *Homestead* 49 (1904): 896. *Iowa Yearbook of Agriculture* 5 (1904): 130. See also *Homestead* 49 (1904): 947; "Exit the State Fair Faker," *Homestead* 50 (1905): 3; *Wallaces' Farmer* 30 (1904): 1041; "Why Not Muzzle the State Fair Barkers?," *Wallaces' Farmer* 30 (1904): 1062.

20. *Wallaces' Farmer* 29 (1904): 1041. See also "The State Fair," *Wallaces' Farmer* 29 (1904): 1066.

21. *Iowa Yearbook of Agriculture* 5 (1904): 98; 6 (1905): 97.

22. JCS to Charles Downing, May 16, 1904, State Board of Agriculture papers (henceforth SBA papers), State Archives of Iowa, Des Moines; JCS to W. D. Ament, May 1906, SBA papers, box Sat 3, f. 27; Ament to JCS, May 16, 1906, SBA papers, f. 27; JCS to W. W. Wheeler (Shelby County Fair Association), April 19, 1907, SBA papers, box Sz 10, f. 86. Advertisements for a diverse array of acts may be found in series AD V, boxes Sat 1, 2, 5, 6.

23. *Iowa Yearbook of Agriculture* 6 (1905): 160; 7 (1906): 144; JCS to W. W. Wheeler (Shelby County Fair Association), April 19, 1907, SBA papers, box Sz 10, f. 86; *Billboard* 27, no. 14 (April 3, 1915): 3.

24. *The Des Moines Register and Leader,* August 21, 1908. For a similar description of the Midway at the 1910 fair, see August 28, 1910.

25. "The Men Who Make Fairs Successful," *Billboard* 21, no. 49 (December 4, 1909): 14.

26. W. D. Ament to JCS, May 9, May 15, 1905, SBA papers, Sat 1, f. 12.

27. In 1910 the Showmen's Protective Association was formed. Its object, according to *Billboard*, was "to drive from the business all illegitimate workers." "Shall Fair Showmen Organize?" *Billboard* 22, no. 9 (February 26, 1910): 12; see also "The Concern of Showmen," *Billboard* 22, no. 43 (October 22, 1910): 3.

28. "Side Shows and Concessions," *Billboard* 21, no. 11 (March 13, 1909): 16; see also "Attractions at Fairs," *Billboard* 21, no. 16 (April 17, 1909): 16; 21, no. 7 (February 13, 1909): 16; "The New Era of Carnivaldom," *Billboard* 26, no. 51 (December 19, 1914): 48. J. W. Fleming (Assistant Secretary, Ohio State Fair), "System in Fair Management Solution of the Problem That Has Vexed Many Associations," *Billboard* 20, no. 49 (December 5, 1908): 14; Fleming, "Scientific Management of Fairs," *Billboard* 23, no. 25 (June 24, 1911): 8. See also "How Modern Business Methods Can Be Applied By Carnival Sheiks," *Billboard* 26, no. 12 (March 21, 1914): 34; Don V. Moore, "Keeping Carnivals Clean," *Country Gentleman* 90 (May 30, 1925): 17.

29. "Attractions at Fairs," *Billboard* 21, no. 16 (April 17, 1909): 16; "Amusement Features of Fairs, *Billboard* 22, no. 3 (January 15, 1910): 14.

30. On fair men's efforts to band together to preserve their industry, see A. L. Denio to JCS, March 1908, SBA papers, box Sz 11, f. 92; *Iowa Yearbook of Agriculture* 17 (1916): 137–39; 20 (1919): 178; 23 (1922): 65–66; Iowa Fair Managers Association convention, *Iowa Yearbook of Agriculture* 14 (1913): 243, 254–55.

31. *Des Moines Leader*, August 29, 1900.

32. JCS to I. S. Mahan (Assistant Secretary, Oklahoma State Fair Association), March 26, 1908, SBA papers, box Sat 5, f. 50; "Amusements at Our Fairs," *Homestead* 55 (1910): 254.

33. *Wallaces' Farmer* 37 (1912): 1256–57. *Homestead* 56 (1911): 1571–72; 61 (1916): 1590; see also *Homestead* 60 (1915): 1507–8; *Wallaces' Farmer* 39 (1914): 1234.

34. *The Des Moines Register and Leader*, August 20, 1911.

35. James Pierce, "The State Fair This Year," *Homestead* 64 (1919): 2043.

36. "Amusement Features at Fairs," *Billboard* 22 no. 3 (January, 15, 1910): 14.

37. "Evolution, Not Revolution," 22.

38. *Iowa Yearbook of Agriculture* 16 (1915): 149.

39. State Board of Agriculture, *Official Catalog*, 1919: 23–24. For more examples of the fair's extensive entertainment offerings and expenditures in the *Official Catalog*, see 1913: 30–32; 1914: 65–66; 1924: 25–28; 1925: 30; 1926: 23–26; 1927: 27–28; *Iowa Yearbook of Agriculture* 21 (1920): 9; 22 (1921); 11.

40. For notable examples of the resurgence of agrarian rhetoric in the 1920s,

see "The World's Greatest State Fair," *Homestead* 66 (1921): 1341; Iowa State Fair Board, *Premium List*, 1922: 7. On county fairs' waning popularity, see Walter E. Olson, "Why County Fairs Are Sick," *Billboard* 44, no. 13 (March 26, 1932): 4; F. J. Claypool, "Have County Fairs Seen Their Day?" *Billboard* 44, no. 24 (June 11, 1932): 64. In 1925, Iowa State Fair secretary Arthur Corey observed that seventy-four Iowa county fairs had made money during the past year, while twenty-three had lost money. Without assistance from the state government, however, eighty of the state's ninety-seven county fairs would have been in the red. *Homestead* 70 (1925): 1778. "Better County Fairs for Iowa," *Homestead* 66 (1921): 2173. On the state fair's growing entertainment budget, see, for instance, *Iowa Yearbook of Agriculture* 22 (1921): 11, which records that the fair board paid more than $40,000 to book acts for the 1921 fair.

41. Carl N. Kennedy, "Will Our Local Fairs Survive?" *Wallaces' Farmer and Iowa Homestead* 55 (1930): 1917. In 1925, Iowa State Fair secretary Arthur Corey observed that seventy-four Iowa county fairs had made money during the past year, while twenty-three had lost money. Without assistance from the state government, however, eighty of the state's ninety-seven county fairs would have been in the red. *Homestead* 70 (1925): 1778. "Better County Fairs for Iowa," *Homestead* 66 (1921): 2173. On the state fair's growing entertainment budget, see, for instance, *Iowa Yearbook of Agriculture* 22 (1921): 11, which records that the fair board paid more than $40,000 to book acts for the 1921 fair. *Iowa Yearbook of Agriculture* 22 (1921): 205.

42. On fair men's efforts to band together to preserve their industry, see A. L. Denio to JCS, March 1908, SBA papers, box Sz 11, f. 92; *Iowa Yearbook of Agriculture* 17 (1916): 137–39; 20 (1919): 178; 23 (1922): 65–66; Iowa Fair Managers Association convention, *Iowa Yearbook of Agriculture* 14 (1913): 243, 254–55.

43. I. S. Bailey (Grinnell Fair Association), "Fairs and Expositions," *Iowa Yearbook of Agriculture* 21 (1920): 156–60; Carl N. Kennedy, "Will Our Local Fairs Survive?" *Billboard* 55 (1930): 1935. *Iowa Yearbook of Agriculture* 22 (1921): 205; "Confessions of a Fair Faker," *Country Gentleman* 87 (April 8, 1922): 4–5; 87 (April 15, 1922): 7; 87 (April 22, 1922): 9. *Country Gentleman* was published in Philadelphia but circulated nationally and addressed topics of interest to farmers throughout the United States. See also Courtney R. Cooper, "The Fixer and the Fake," *Country Gentleman* 87 (May 6, 1922): 7, 22.

44. "The Uprising Against the Fakes," *Country Gentleman* 87 (June 3, 1922): 6; *Iowa Yearbook of Agriculture* 23 (1922): 47–53, 67. Moore literally compiled a scrapbook of articles about crooked carnivals, which he took to the annual convention of the International Association of Fairs and Expositions in November 1922; there he persuaded the association to adopt a resolution in favor of "clean fairs." A. B. Macdonald, "The Scrapbook of Fair Fakery," *Country Gentleman* 88 (February 17, 1923): 1–2, 36.

45. "The Scrapbook of Fair Fakery," 1; Dante M. Pierce, "Cut Out the Fakes at the Fairs," *Homestead* 68 (1923): 1177.

46. John C. Simpson, addressing the Iowa Fair Managers Convention, quoted in *Greater Iowa*, December 1923: 6.

47. A. B. Macdonald, "It's Now or Never for the Carnivals," *Country Gentleman* 88 (June 2, 1923): 3–4; see, in the same issue, the editorial "The Carnival's Reprieve," 14; Thomas J. Johnson, "Crooked Carnivals," *Country Gentleman* 90 (April 25, 1925): 7; Macdonald, "Carnivals Must Clean Up Or Be Cleaned," *Country Gentleman* 88 (December 1, 1923): 8. On the uphill battle to clean up the carnival business, see A. B. Macdonald, "The Nickel Nicker, the Gimmick and the Yap," *Country Gentleman* 91 (May 1926): 9, 92; C. G. Dodson, "Carnivals, Committees and Fairs," *Billboard* 39, no. 50 (December 10, 1927): 87.

48. A. B. Macdonald, "Hop-Scotch Grifters," *Country Gentleman* 89 (May 10, 1924): 15, 31.

49. For *Billboard's* defense of the carnival business, see William J. Hilliar, "The Carnival and Its 'Working Men,'" *Billboard* 38, no. 12 (March 20, 1926): 11; see also Fred Beckmann (of Beckmann and Gerety Shows), "The Carnival—Its Inception, Its Progress, Its Status," *Billboard* 37, no. 50 (December 12, 1925): 14.

50. Macdonald, "Hop-Scotch Grifters," 15, 31; "Never Were Fairs Better or Cleaner," *Homestead* 68 (1923): 1502; Dante M. Pierce, "Small Fairs Have Gained Ground in Iowa," *Homestead* 70 (1925): 1747; "Fairs Continue in Public Favor; Public Believes in Annual Event," *Billboard* 40, no. 48 (December 1, 1928): 66. See also "J. C. Simpson Sees Prosperous Year," *Billboard* 37, no. 10 (March 7, 1925): 81; Robert E. Hickey, "Fair Attractions and Fair Attendance," *Billboard* 37, no. 12 (March 21, 1925): 16–17.

51. *The Des Moines Register*, August 28, 1921; Harlan Miller, "Midway Wonders Attract Fools and Wise Men Alike," *The Des Moines Register*, August 25, 1929.

52. James M. Pierce, "Can Moving Pictures Be Brought to the Farm?" *Homestead* 65 (1920): 2386.

53. JCS to I. S. Mahan (Assistant Secretary, Oklahoma State Fair Association), March 26, 1908, SBA papers, box Sat 5, f. 50. On *Billboard's* efforts to preserve cooperation among branches of the show business, see "Live and Let Live—Or the Case for Outdoor Amusements," *Billboard* 38, no. 24 (June 5, 1926): 45; Carleton Collins, "Carnivals Leave Money in Town—Instead of Taking Money Away," *Billboard* 38, no. 25 (June 12, 1926): 49; 38, no. 25 (June 19, 1926): 44; 38, no. 27 (July 3, 1926): 44–45; "Carnivals and 'Taking Money Out of Town'—A Challenge," *Billboard* 43, no. 19 (May 9, 1931): 34. The magazine soon became more defensive of the outdoor entertainment industry. See "Outdoor Showmen Must Fight Their Oppressors," *Billboard* 42, no. 18 (May 3, 1930): 44. Despite their bravado on behalf of live performers, *Billboard's* editors could see the final curtain beginning to fall. See "The Last Stand of the Living?" *Billboard* 44, no. 47 (November 19, 1932): 24.

54. "Attractions at Fairs," *Billboard* 21, no. 16 (April 17, 1909): 16; *Breeder's Gazette*, quoted in *Iowa Yearbook of Agriculture* 10 (1909): 660.

55. JCS to James W. Fleming (Assistant Secretary, Ohio State Fair), March 8, 1909, SBA papers, box Sz 13, f. 116; *Breeder's Gazette*, quoted in *Iowa Yearbook of Agriculture* 10 (1909): 661; L. R. Fairall, *Iowa Yearbook of Agriculture* 23 (1922): 184; J. C. McCaffrey (of the Western Vaudeville Managers Association), "Attractions at Fairs and Parks," *Billboard* 38 no. 12 (March 20, 1926): 18; McCaffrey, "The Fair Pageant Has Arrived," *Billboard* 38, no. 14 (April 3, 1926): 45. On the growing popularity of more spectacular, passive entertainments, see Neil Harris, *Humbug*, and Lawrence Levine, *Highbrow/Lowbrow* (Cambridge, MA: Harvard University Press, 1988).

56. Mrs. Eugene Cutler, "A Fair for All," Iowa State Fair Board, *Annual Report*, 1938: 177; *The Des Moines Register*, August 27, 1927.

57. David Glassberg, *American Historical Pageantry: The Uses of Tradition in the Early Twentieth Century* (Chapel Hill: University of North Carolina Press, 1990).

58. On these annual fireworks finales, see State Board of Agriculture, *Official Catalog*, 1914: 92–95.

59. B. E. Gregory (of the Pain Manufacturing Company) to JCS, May 14, 1902, SBA papers, box Sat 1, f. 7. *The Des Moines Register and Leader*, August 21, 25, 27, 1902, August 22, 25, 1903. The fair also booked a historical spectacle, "Battle of San Juan," in 1899. *Des Moines Leader* August 26, 27, 1899.

60. JCS to W. W. Morrow, May 22, 1905, JCS to C. E. Cameron, May 22, 1905, Morrow to JCS, May 23, 1905, SBA papers, box Sat 3, f. 21; *The Des Moines Register and Leader*, August 28, 1905; *Iowa Yearbook of Agriculture* 6 (1905): 156, 159; 15 (1914): 44; Iowa State Fair Board, *Official Catalog*, 1914: 92–95; *The Des Moines Register and Leader*, August 28, 29, 1914.

61. *Iowa Yearbook of Agriculture* 16 (1915): 8; *Greater Iowa*, August 1915; *The Des Moines Register and Leader*, August 23, 26, 1915; Iowa State Fair Board, *Official Catalog*, 1915: 38. On "Modern Warfare," see State Board of Agriculture, *Official Catalog*, 1917: 35; *The Des Moines Register*, August 25, 1917; *Greater Iowa*, September 1917: 2; *Iowa Yearbook of Agriculture* 18 (1917): 76–77. "World's War" is described in *The Des Moines Register*, August 24, 1918; see also *Iowa Yearbook of Agriculture* 19 (1918): 10; 20 (1919): 13; *Greater Iowa*, March 1918: 8. The following year, "The Battle of Chateau Thierry" re-created the decisive, but costly, Allied victory. In 1920 the fair re-created one of the Allies' naval battles, "Siege of the Dardanelles." *The Des Moines Register*, August 20, 23, 24, 1919; *Homestead* 64 (1919): 1956, 1972. *The Des Moines Register*, August 28, 1920.

62. *Des Moines Register and Leader*, August 23, 1913, Fair magazine section, p. 6; Iowa State Fair Board, *Official Catalog*, 1913: 30, 33–35, 46–49; *Greater Iowa*, August 15, 1913: 8. Also, in 1908 the fair staged "Sheridan's Ride," a Civil War tableau. *Greater Iowa*, May 1928: 5; *The Des Moines Register*, August 24, 25, 1928. Only once did the fair's spectacle re-create an event of Iowa history, when, in 1910, the Pain Company staged a dramatization of "The Spirit Lake Massacre of 1857," in which a band of Sioux Indians killed thirty-four settlers. *Iowa Yearbook*

of Agriculture 11 (1910): 301, 303; *Greater Iowa*, May 1910: 1; August 12, 1910: 2; *Des Moines Register and Leader*, August 21, 1910, Exposition number, p. 5.

63. On Rome, see *Greater Iowa*, April 1925: 3; June 1925: 5; *The Des Moines Register*, August 27, 28, 1925; on ancient Greece, see *Greater Iowa*, April 1927: 2; Iowa State Fair Board, *Official Catalog*, 1927: 25; *The Des Moines Register*, August 19, 27, 1927. On "Montezuma," see *Greater Iowa*, June 1921: 8; *Homestead* 66 (1921): 353.

64. On the several productions of "Pompeii," see *Greater Iowa*, August 19, 1911: 6; *Des Moines Register and Leader*, August 20, 1911, Fair magazine section, p. 3; August 30, 1911. *Iowa Yearbook of Agriculture* 17 (1916): executive committee meeting, March 6–7, 1916; *Greater Iowa*, July 1916: 5; State Board of Agriculture, *Official Catalog*, 1916: 35; *The Des Moines Register*, August 20, 1916; *Homestead* 61 (1916): 1418.

65. *Greater Iowa*, April 1929: 7; July 1929: 2; *The Des Moines Register*, August 24, 1929; *Wallaces' Farmer* 54 (1929): 1133; *Homestead* 74 (1929): 1336. At least one other spectacle of volcanic destruction, "Mt. Pelee and the Destruction of St. Pierre," graced the fairgrounds. See *Homestead* 49 (1904): 1082. *Greater Iowa*, May 1924: 7; *The Des Moines Register*, August 21, 23, 1924.

66. F. M. Barnes to JCS, April 14, 1909, SBA papers, box Sat 6, f. 56; JCS to C. E. Cameron, April 15, 1909, Cameron to JCS, April 19, SBA papers, box Sat 6, f. 58. *Greater Iowa*, May 1909: 1; August 14, 1909: 6; Iowa State Fair Board, *Official Catalog*, 1909: 18; *Iowa Yearbook of Agriculture* 10 (1909): 228; *Des Moines Register and Leader*, August 20, 29, 1909.

67. On "Mystic China," see *Iowa Yearbook of Agriculture* 23 (1922): 2; State Board of Agriculture, *Official Catalog*, 1922: 61; *Greater Iowa*, July 1922. "India" is described in *Greater Iowa*, May 1923: 6; *The Des Moines Register*, August 24, 25, 1923. "Baghdad" is reported in *Greater Iowa*, May 1928: 5; *The Des Moines Register*, August 24, 25, 1928. On the telling shift in pageant themes in the 1920s, see Glassberg, *American Historical Pageantry*, 5.

68. *Greater Iowa*, August 1930: 4; Iowa State Fair Board, *Official Catalog*, 1930: 22–23; *The Des Moines Register*, August 23, 1930.

69. In 1931 and 1932 the fair booked variety shows that were an odd mixture of music and travelogue, affording the audience glimpses of various cities, cultures, and festivities around the globe. *Greater Iowa*, May 1931: 4; August 1931: 4; Iowa State Fair Board, *Official Catalog*, 1931: 22–24; *The Des Moines Register*, August 30, 193l; *Greater Iowa*, June 1932: 5; Iowa State Fair Board, *Official Catalog*, 1932: 21–23; *The Des Moines Register*, August 26, 27, 1932.

70. F. A. Schmidt, *Train Wrecks for Fun and Profit* (Erin, Ontario: Boston Mills Press, 1982), 42–45; *Greater Iowa*, August 1932: 1, 3; poster advertisement for collision, Manuscript BL Io9, SHSI, Des Moines. James J. Reisdorff, *The Man Who Wrecked 146 Locomotives: The Story of "Head-On Joe" Connolly* (David City, NE: South Platte Press, 2009).

71. *The Des Moines Register*, August 28, August 29, 1932.

72. In 1931 the fair lost more than $63,000; the following year, it lost $67,000. Iowa State Fair Board, *Premium List*, 1931: 7. Iowa State Fair Board, *Official Catalog*, 1932: 9; *Greater Iowa*, April 1932: 2; May 1933: 2; "Keep the Fairs Going," *Billboard* 44, no. 53 (December 31, 1932): 58.

73. "Spare the Fairs Their State Aid," *Billboard* 45, no. 7 (February 18, 1933): 24.

74. "Lack of Showmanship Chief Reason Behind Fair Failures," *Billboard* 42, no. 14 (April 5, 1930): 46; "Fairs Need Bigger and Better Professional Entertainment," *Billboard* 43, no. 3 (January 17, 1931): 44. *Billboard* published many editorials urging fair men not to skimp on entertainments. See, for instance, "Professional Entertainment and What It Means to a Fair," *Billboard* 42, no. 37 (September 13, 1930): 44; "Carnivals Offer Great Possibilities to Those Who Will Pioneer in Ideas," *Billboard* 43, no. 2 (January 10, 1931): 44; "Conditions in Fair Business Not as Bad as Calamity Howlers Say," *Billboard* 43, no. 33 (August 15, 1931): 28; "Going to Extremes on Economizing on Fair Grand-Stand Entertainment," *Billboard* 44, no. 3 (January 16, 1932): 28.

Chapter 5

1. Ruth Suckow, "Iowa," *American Mercury* 1926, 39–45. For an account of regionalist thought in Iowa, see E. Bradford Burns, *Kinship with the Land: Regionalist Thought in Iowa, 1894–1942* (Iowa City: University of Iowa Press, 1996).

2. F. Scott Fitzgerald, *The Great Gatsby* (New York: Scribner's, 1925), 182.

3. On Stong, see Clarence A. Andrews, *A Literary History of Iowa* (Iowa City: University of Iowa Press, 1972), 103–13. The most thorough analysis of Henry King's film is found in Walter Coppedge, *Henry King's America* (Metuchen, NJ: Scarecrow Press, 1986), 72–91.

4. George C. Duffield, *Memories of Frontier Iowa* (Des Moines: Bishard Brothers, 1906).

5. Philip Duffield Stong (henceforth PDS) to Ben L. Stong, June 27, 1918, Philip D. Stong papers, Archives, Cowles Library, Drake University, Des Moines. Except where otherwise noted, all references to Stong's correspondence are from this collection.

6. PDS to Folks, April 18, 1921.

7. PDS to Folks, February 11, February 15, July 5, July 11, 1925, February 28, 1927, May 16, 1927, January 20, 1929.

8. PDS to Folks, June 6, July 28, 1931. For Stong's recollections of writing the novel, see PDS to Cyril Clemens, February 23, 1956, Department of Special Collections, University of Iowa Library.

9. PDS to Folks, November 19, 1931, December 21, 1931.

10. PDS to Folks, January 10, 1932.

11. PDS, telegram to Mrs. B. J. Stong, June 8, 1932; PDS to Folks, May 16, 1932.

12. PDS to Folks, June 17, 1932; PDS, telegram to Ben Stong, June 24, 1932; PDS to Folks, June 26, 1932, November 22, 1932; *New York Times*, February 26,1933; Stong, *Hawkeyes*, 94–95. Stong gradually used his windfall to purchase Linwood, the farm his grandfather had once owned, and he subsequently bought several adjoining parcels of land in order to reassemble the original Duffield farm. PDS to Folks, September 5, 13, October 30, 1932, January 22, 1933, August 24, 1933.

13. Phil Stong, *State Fair* (Philadelphia: Century, 1932), 9. See also PDS to Folks, September 12, 1928.

14. *State Fair*, 9–12.

15. *State Fair*, 13–14.

16. *State Fair*, 17–22, 28–29.

17. *State Fair*, 53, 54, 62–63.

18. *State Fair*, 97–98.

19. *State Fair*, 86, 98, 87, 104, 146–47, 155–58.

20. *State Fair*, 217, 219.

21. *State Fair*, 237.

22. *State Fair*, 244, 252–53.

23. *State Fair*, 254, 263.

24. Donald R. Murphy, "Sedate Saturnalia, Sober Bacchanals," *The Des Moines Register*, May 22, 1932. Stong was understandably irked by the *Register*'s critical review. See PDS to Folks, May 16, 1932. Louis Kronenberger, "The Brighter Side of Farm Life," *New York Times Book Review*, May 8,1932: 6; Robert Cantwell, "This Side of Paradise," *New Republic* 71 (July 6, 1932): 215–16.

25. John L. Shover, *Corn Belt Rebellion: The Farmers Holiday Association* (Urbana: University of Illinois Press, 1965).

26. PDS to Folks, July 18, 1932.

27. PDS to Folks, August 7, 1932. See also PDS to Folks, July 25, 1932.

28. PDS to Folks, February 17, 1933.

29. Henry King, Oral History, Directors Guild of America, "State Fair" folder, Film Studies Department, Museum of Modern Art, New York. See also Coppedge, *Henry King's America*. In the film, Emily is transformed from the daughter of a horseman into a trapeze artist.

30. Eugene Burr, "State Fair," *Billboard* 45, no. 5 (February 4, 1933): 12; *Variety*, January 31, 1933: 12; *The Des Moines Register*, February 18, 1933.

31. MacDonald's review appeared in *The Symposium*, April and July 1933, and is

excerpted in *Dwight Macdonald on Movies* (Englewood Cliffs, NJ: Prentice-Hall, 1969), 88.

32. On Fox's narrow escape from receivership, see *Billboard* 45, no. 14 (April 8, 1933): 4; Coppedge, *Henry King's America*, 73, 86; *Time*, January 29, 1934: 41.

33. PDS to Folks, March 8, 1932.

34. Phil Stong, *Stranger's Return* (New York: Harcourt, Brace, 1933); Stong, *Village Tale* (New York: Harcourt, Brace, 1934). Louis Kronenberger, "Phil Stong's Village Tale," *New York Times Book Review*, March 11, 1934: 8.

35. PDS to L. F. Fairall (Director of Publicity for the Iowa State Fair), April 30, 1934, Department of Special Collections, University of Iowa Library.

36. PDS to Folks, June 2, 1934. In November, however, RKO bought rights to the novel for $10,000. PDS to Folks, November 7, 1934. Phil Stong, *Week-End* (New York: Harcourt, Brace, 1935), 245, 261–67.

37. PDS to Folks, February 15, 1936.

38. Stong's subsequent Iowa novels were *Buckskin Breeches* (1937), *The Long Lane* (1939), *The Princess* (1941), *One Destiny* (1942), *Return in August* (1953), and *Blizzard* (1955). Stong also published a history of Iowa: Phil Stong, *Hawkeyes: A Biography of the State of Iowa* (New York: Dodd, Mead, 1940).

39. The enduring appeal of *State Fair* was confirmed in 1945, when Fox released the Rodgers and Hammerstein musical based on the novel, which became the basis for a second film version of the musical in 1962, starring Pat Boone and Ann-Margret, and the story has been presented on Broadway as recently as 1996. Phil Stong, *Return in August* (Garden City, New York: Doubleday, 1953).

40. For similarly harsh assessments of Stong's work, see Ronald Weber, *The Midwestern Ascendancy in American Writing* (Bloomington: University of Indiana Press, 1992), 191–95; Roy W. Meyer, *The Middle-Western Farm Novel in the Twentieth Century* (Lincoln: University of Nebraska Press, 1965), 98–99.

41. ISAS, *Report* 1 (1854): 18. See the "List of Premiums" published in the *Iowa Farmer and Horticulturalist* 2 (1854): 60. See also ISAS, *Report* 1 (1854): 18.

42. See A. B. Miller to JRS, July 28, 1883, ISAS papers, box Sz 1, f. 4.

43. "History of the Fair," *Northwest Farmer*, quoted in ISAS, *Report* 6 (1859): 77; *Iowa Homestead and Western Farm Journal* 15 (23 September 1870): 4. See also *Iowa State Register*, September 10, 1886, which asserted that "that State Fair is most undoubtedly the place where artists should show their pictures in order that the progress of the State may be noted by the people of the State when this annual gathering of thousands takes place."

44. *Iowa State Register*, September 8, 1882.

45. *North-West Farmer*, quoted in ISAS, *Report* 3 (1856): 29; ISAS, *Report* 16 (1869): 167. N. B. Collins to Mrs. L. B. James, September 20, 1889, ISAS papers, box Se 2, f. 6.

46. JMS to R. F. Bower, October 1, 1870. Also JRS to Annette Butler, November 4, 1882; JRS to A. B. Miller, August 1, 1883.

47. See, for instance, JRS to E. R. Shankland, November 3, 1882; JRS to A. B. Miller, August 1, 1883.

48. *Western Farm Journal* 18 (September 26, 1873): [1].

49. *Iowa State Register*, September 4, 1879: [4]. The following year, the Register contended that all the fair's displays were meritorious, "except that of fine arts in the stricter sense of art." *Iowa State Register*, September 11, 1880; *Iowa Homestead* 27 (September 15, 1882): 4. See also *Iowa State Register*, September 8, 1887.

50. *Iowa State Register*, September 4, 1890.

51. *Iowa State Register*, September 2, 1891, September 1, 1892.

52. See, for example, Mrs. H. Perrior to JRS, September 1889, ISAS papers, box Se 2, f. 9; Mrs. P. L. Fields to JRS, September 16, 1889, ISAS papers, box Se 2, f. 7; JRS to W. W. Field, September 21, 1893; E. Everson to P. L. Fowler, August 22, 1895, ISAS papers, box Sz 2, f. 19.

53. "Art in Iowa," *Des Moines Leader*, September 13, 1896.

54. "Art in Iowa," *Des Moines Leader*, September 13, 1896.

55. Bess Ferguson, with Velma Rayness and Edna Gouwens, *Charles Atherton Cumming: Iowa's Pioneer Artist-Educator* (Des Moines: Iowa Art Guild, 1972). Cumming headed the Des Moines Academy of Art from 1895 to 1900, when he founded the Cumming School of Art.

56. *Des Moines Leader*, September 18, 1897.

57. On the creation of the Fine Arts Hall, see *Des Moines Leader*, August 26, 1899; ISAS, *Report* 46 (1899): 22, 28, 99; G. H. Van Houten to L. H. Pickard, February 6, 1899; GHV to LHP, March 2, 1899. Cumming scrapbook, Iowa Art Guild, Des Moines, 13.

58. On the disagreement between Cumming and the fair board, see the *Des Moines Leader*, August 26, 1899. On Cumming's efforts to improve the quality of the art exhibit, see *Iowa Yearbook of Agriculture* 15 (1914): 7, 31; *Greater Iowa*, May 1914: 8. See also the unflattering report of T. C. Legoe, director of the Fine Arts Department, to the State Board of Agriculture in 1909. *Iowa Yearbook of Agriculture* 10 (1909): 276–77.

59. *Des Moines Register and Leader*, September 1, 1914. *Greater Iowa*, May 1914: 8; August 1914: 5.

60. Iowa State Board of Agriculture, *Official Catalog*, 1916: 23.

61. In addition to assuming control of the state fair's art exhibit, Cumming also presided over the founding of the Iowa Art Guild, an organization of conservative academic painters, in September, 1914. *Greater Iowa*, May 1914: 6.

62. See Cumming's inflammatory and anti-Semitic letter to University of Iowa President W. A. Jessup, May 22, 1924, Charles Atherton Cumming papers, Department of Special Collections, University of Iowa Library.

63. *The Des Moines Register*, August 27, 1925: 1; August 28, 1925: 7. Cumming's dismissal of Mrs. Greenman generated considerable controversy. For a defense of Cumming's actions, see Lewis Worthington Smith's newspaper editorial, "Judgment in Art," in Ferguson, *Charles Atherton Cumming*, 46.

64. *The Des Moines Register*, September 4, 1925.

65. Cumming authored a pamphlet, *A Defense of the White Man's Art*, a blueprint for a much longer work that remained uncompleted. Charles Atherton Cumming papers, Department of Manuscripts, State Historical Society of Iowa, Des Moines. See also Cumming to Benjamin F. Shambaugh, June 30, 1928, Charles Atherton Cumming papers, Department of Special Collections, University of Iowa Library. In 1932 Cumming returned to Iowa, where he died shortly after his arrival. His obituaries may be found in both *The Des Moines Register* and *Des Moines Tribune*, February 18, 1932.

66. *Greater Iowa*, July 1927: 7.

67. For two differing accounts of Wood's conversion to regionalism, see James M. Dennis, *Grant Wood: A Study in American Art and Culture* (New York: Viking, 1975), and Wanda Corn, *Grant Wood: The Regionalist Vision* (New Haven, CT: Yale University Press, 1983), 25–33. See also James M. Dennis, *Renegade Regionalists: The Modern Independence of Grant Wood, Thomas Hart Benton, and John Steuart Curry* (Madison: University of Wisconsin Press, 1998), 173–96; R. Tripp Evans, *Grant Wood (A Life)* (New York: Alfred A. Knopf, 2010).

68. Wood later retitled *Portrait of Arnold Pyle*, renaming it *Arnold Comes of Age*. Wood to Zenobia B. Ness, October 28, 1930, Archives of American Art. He exhibited *Dinner for Threshers* at the fair in 1933.

69. On the Stone City colony, see Corn, *Grant Wood*, 38–43; Dennis, *Grant Wood*, 195; Grant Wood, "Aim of the Colony," in the pamphlet *Stone City* pamphlet (1933). Iowa Artists scrapbook, State Historical Society of Iowa, Des Moines; Frederick A. Whiting Jr., "Stone, Steel, and Fire: Stone City Comes to Life," *American Magazine of Art* 25 (1932): 333–42.

70. Grant Wood, *Revolt Against the City* (Iowa City: Clio Press, 1935), reprinted in Dennis, *Grant Wood*, 229–35 (quotation from p. 231); Mott's authorship of the essay is discussed in Corn, *Grant Wood*, 46, 153.

71. On Wood's influences, see Dennis, *Grant Wood*, esp. 67–8, 197. As Dennis notes, "the term 'regionalism' knows no equal for ambiguity." Dennis, *Grant Wood*, 143.

72. Craven published several paeans to regionalism, including his *Modern Art* (New York: Simon and Schuster, 1934), esp. 260–72, 320–31, and *Men of Art*, (New York: Simon and Schuster, 1934), 506–13. See also the flattering assessment

of regionalism, "U.S. Scene," *Time*, December 24, 1934. Davis's broadside appeared in *Art Front* (February 1935) and is reprinted in Diane Kelder, ed., *Stuart Davis* (New York: Praeger, 1971), 151–54.

73. Harnoncourt later became director of the Museum of Modern Art. Dornbusch won the prize with *The Road Ahead*, and Pyle won for *Sunday Morning*. *The Des Moines Register*, August 24, 1933; "State Fair Art—Indigenous Fertility," *American Magazine of Art* 26 (1933): 476–77.

74. *The Des Moines Register*, August 22, 1934.

75. See *The Des Moines Register*, August 22, 1934, August 18, 21, 1935; see also *Greater Iowa*, July 1935: 3; Joseph S. Czestochowski, *Marvin Cone: An American Tradition* (New York: E. P. Dutton, 1985).

76. *Des Moines Register*, August 22, 1935.

77. Iowa State Fair Board, *Premium List*, 1935: 185; *Greater Iowa*, July 1935: 3; *The Des Moines Register*, September 1, 1935.

78. The Co-operative Artists, like many leftist organizations, soon fell victim to internecine strife. Heated disputes over its political orientation led to the union's dissolution in 1937, and White and many of his comrades returned to the state fair's competition. *The Des Moines Register*, August 26, 30, 1936, August 25, 1937. Much has been written on the similarities and differences between regionalism and Social Realism. The best place to begin is Matthew Baigell, *The American Scene: American Painting of the 1930s* (New York: Praeger, 1974), 55–73, 74.

79. *The Des Moines Register*, August 25, 1937; *Wallaces' Farmer*, quoted in *Greater Iowa*, October 1937: 4.

80. Daniel Rhodes, telephone conversation with author, December 14, 1988; *The Des Moines Register*, August 24, 1938.

81. *Wallaces' Farmer*, quoted in Iowa State Fair Board, *Annual Report*, 1938: 18.

82. *Wallaces' Farmer* 63 (September 10, 1938): 592.

83. Rhodes and Johnson each painted half of this mural, with Rhodes executing the earlier years of Iowa's history and Johnson the more recent decades. For two engaging examinations of the response of ordinary Americans to regionalist painting, see Karal Ann Marling, *Wall-to Wall America: Post Office Murals in the Great Depression* (Minneapolis: University of Minnesota Press, 1982); M. Sue Kendall, *Rethinking Regionalism: John Steuart Curry and the Kansas Mural Controversy* (Washington, DC: Smithsonian Institution Press, 1986).

84. *The Des Moines Register*, August 24, 1939: 3; ISAS, *Report* 1 (1854): 28.

85. *The Des Moines Register*, August 23, 1939, August 21, 1940. Later, Rhodes moved to California, where he became a renowned ceramicist.

86. Gebers, a part-time schoolteacher, had studied under Wood at Stone City and chaired the art division of the Iowa Federation of Women's Clubs; Eliza-

beth Wherry, "Iowa Art Is Never Tiresome," *Wallaces' Farmer* 66 (September 6, 1941): 576. Grant Wood himself died in February 1942. For an exploration of the demise of regionalism and advent of abstract expressionism, see Erika Doss, *Benton, Pollock and the Politics of Modernism: From Regionalism to Abstract Expressionism* (Chicago: University of Chicago Press, 1991), chaps. 4–5.

87. *The Des Moines Register,* June 25, 1946.

Conclusion

1. ISAS, *Report* 1 (1854): 28. See A. P. Sandles (Secretary, Ohio State Fair), "The Origin of Fairs," *Billboard* 23, no. 31 (August 5, 1911): 3; H. J. Aymer, "The Evolution of the Fair," 24, no. 12 (March 23, 1912): 19; James M. Chaffaut (Publicity Director, Ohio State Fair), "The Changing Fair," 43, no. 13 (March 28, 1931): 11. See also Neely, *The Agricultural Fair,* esp. 251–64.

2. On the fair's increasingly backward-looking exhibits in the 1890s, see G. H. Van Houten to L. H. Pickard, March 7, 1899, ISAS papers, AD V, book 54:122–23; *Homestead* 54 (1909): 1402–3. On the historical aspects of the 1904 fair, see the special Fair section in *The Des Moines Register and Leader,* August 21, 1904, which includes accounts of the first fair, the fair's subsequent development, and "a prophecy" by State Board of Agriculture member S. B. Packard on "The State Fair of 1954."

3. Frederick Jackson Turner, "The Significance of the Frontier in American History," *Proceedings of the Forty-First Annual Meeting of the State Historical Society of Wisconsin* (Madison, 1894), 79–112; Andrew R. L. Cayton and Peter S. Onuf, *The Midwest and the Nation: Rethinking the History of an American Region* (Bloomington: Indiana University Press, 1989).

4. *Greater Iowa,* April 1929: 7; July 1929: 2; *The Des Moines Register,* August 24, 1929; *Wallaces' Farmer* 54 (1929): 1133; *Homestead* 74 (1929): 1336.

5. On the historical aspects of the "Diamond Jubilee," see *Greater Iowa,* October 1929: 5, 6; *Wallaces' Farmer* 54 (1929): 1189; *Homestead* 74 (1929):1166, 1174, 1209, 1291; *Bureau Farmer* 4 (Iowa edition) (August 1929): 17. Iowans' growing appreciation of their history is lucidly discussed in Ruth Suckow, "Iowa," *American Mercury* 9 (September 1926): 39–45.

6. In 1931 the ravages of the Depression hit with full force, and the fair lost more than $63,000. In 1932 it lost $67,000. Iowa State Fair Board, *Premium List,* 1931: 7. Iowa State Fair Board, *Official Catalog,* 1932: 9; *Greater Iowa,* April 1932: 2; May 1933: 2.

7. Iowa State Fair Board, minutes book, 1937–1940, Iowa State Fairgrounds, Des Moines, 28–30, 70; Iowa Territorial Centennial Committee, "Proposal for Iowa Centennial," 1938, Manuscripts, SHSI, Des Moines, fs. BI.I0278, BI.I0278 cen; Iowa State Fair Board, *Premium List,* 1938: 7. Corey is quoted in Iowa State Fair Board, *Annual Report,* 1938: 5. On the historical pageant, see the souvenir pro-

gram, "Cavalcade of Iowa," SHSI, Des Moines, Manuscripts, f. BI.I0278 (typescript pp. 6–7, 11–13); *Greater Iowa*, August 1938: 5; Iowa State Fair Board, *Official Catalog*, 1938: 43–44; *The Des Moines Register*, August 25, 1938. *Des Moines Tribune*, August 1, 1946; *Wallaces' Farmer* 71 (1946): 645, 691–717.

8. *Wallaces' Farmer and Iowa Homestead* 71 (1946): 715.

9. On nostalgia for an imagined Middle Western "golden age," Shortridge, *The Middle West*, 134–44.